FUNDAMENTALS OF PHYSICS I & II
Lab Manual

AF400648

Charles Cummings

Department of Mathematics and Physics

Grambling State University

All figures, lineart, tables, and photos Courtesy of Charles Cummings.

Cover © Shutterstock.com

www.kendallhunt.com
Send all inquiries to:
4050 Westmark Drive
Dubuque, IA 52004-1840

Copyright © 2018 by Kendall Hunt Publishing Company

ISBN: 978-1-7924-0456-6

All rights reserved. No part of this publication may be reproduced, stored in a retrieval system, or transmitted, in any form or by any means, electronic, mechanical, photocopying, recording, or otherwise, without the prior written permission of the copyright owner.

Published in the United States of America

TABLE OF CONTENTS

LABORATORY 1

Forces Acting on a Single Point

Purpose

To study vectors and to show that the x and y forces acting on a single point in equilibrium cancel or balance each other.

Theory

Forces are vector quantities. The behavior of forces depends on both their magnitudes and directions. Balanced forces must balance in both the x- and y-direction. Each force vector has both x- and y-components.

Consider the force $\mathbf{F}$ diagrammed in Figure 1.1. Its components are given by:

$$F_x = F \cos \theta$$
$$F_y = F \sin \theta$$

where F_x and F_y are the x- and y-components of $\mathbf{F}$, respectively, F is its magnitude, and θ is its angle of direction.

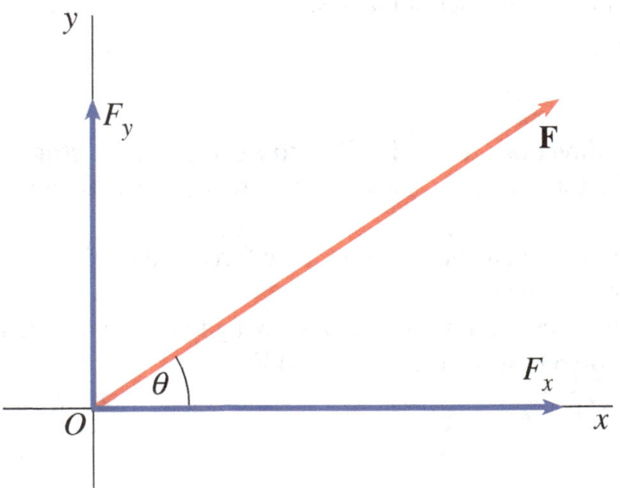

Figure 1.1 Force **F** has x-component F_x parallel to the x-axis and y-component F_y parallel to the y-axis.

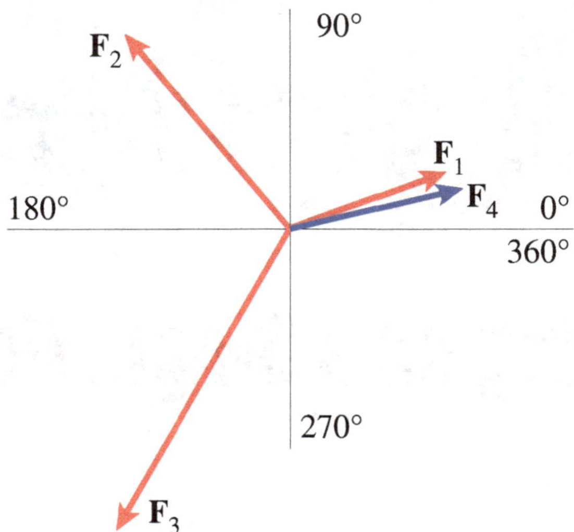

Figure 1.2 Forces **F₁**, **F₂**, and **F₃** are given. Force **F₄** balances the given forces.

Figure 1.2 shows four forces that are in equilibrium. In this experiment, $\mathbf{F}_1$, $\mathbf{F}_2$, and $\mathbf{F}_3$ will be given. $\mathbf{F}_4$ will be the vector which balances the first three. This condition is expressed in vector form as follows:

$$\Sigma \mathbf{F} = 0 = \mathbf{F}_1 + \mathbf{F}_2 + \mathbf{F}_3 + \mathbf{F}_4$$

In component form, the condition for equilibrium is described by two expressions:

$$\Sigma F_x = F_{x1} + F_{x2} + F_{x3} + F_{x4}$$

for the *x*-components and

$$\Sigma F_y = F_{y1} + F_{y2} + F_{y3} + F_{y4}$$

for the *y*-components.

Equipment

Force table, string, weight holders, and set of masses.

Procedure

1. Get the magnitudes and directions forces $\mathbf{F}_1$, $\mathbf{F}_2$, and $\mathbf{F}_3$ from your instructor.
2. Set up each force on the force table as shown in Figure 1.3. Use the correct amount of weight at the correct angle for each.
3. Use the fourth support string to balance the first three forces. In particular, find the angle for which the center stop is at the center of the ring.
4. Add enough weight to balance the system without having to tug on it by hand.
5. Record magnitudes and directions of $\mathbf{F}_1$, $\mathbf{F}_2$, $\mathbf{F}_3$, and $\mathbf{F}_4$.

Data Record

Reproduce Table 1.1 in your laboratory notebook. Record your data in their appropriate locations.

Figure 1.3   Force table with three assigned forces balanced the empirically-determined fourth force.

Data Analysis

The expected result of this experiment is that the net x-component (ΣF_x) and the net y-component (ΣF_y) will both be zero. Due to experimental error, however, this may not be the case. Analysis of data for this experiment includes the determination of ΣF_x and ΣF_y. The accuracy of the experiment is determined by the error in each component of the net force. The errors are given by

$$\%\,\text{error}_x = \frac{2\Sigma F_x}{\Sigma |F_x|}$$

Table 1.1   Force Data

Force	Magnitude (F)	Angle (θ)	F_x	F_y
F_1				
F_2				
F_3				
F_4				

for the *x*-component and

$$\%\text{error}_y = \frac{2\Sigma F_y}{\Sigma |F_y|}$$

for the *y*-component.

Reproduce Table 1.2 in your laboratory notebook. Complete the indicated calculations and analysis.

Table 1.2   Error Analysis

Component	ΣF	$\Sigma |F_i|$ $(i{=}x,y)$	%error
x			
y			

Clothesline

Purpose

To study forces in equilibrium, particularly those acting at small angles such as on a clothesline.

Theory

A small sideways pull can make a very large tension in a rope. Figure 2.1 shows the forces. T_1 and T_2 are the tensions on either end of the rope. F is the sideways pull, and θ_1 and θ_2 are the small angles at which either end of the rope is pulled from a straight line.

The x-components of T_1 and T_2 are $T_1 \cos \theta_1$ and $T_2 \cos \theta_2$, respectively. The y-components are $T_1 \sin \theta_1$ and $T_2 \sin \theta_2$. The forces in both x- and y-directions must be balanced:

$$\Sigma F_x = 0 = T_2 \cos \theta_2 - T_1 \cos \theta_1$$
$$\Sigma F_y = 0 = T_2 \sin \theta_2 + T_1 \sin \theta_1 - F$$

In the case of the sideways force applied to the middle of the rope as shown in Figure 2.2, the deflection angles are equal, $\theta_1 = \theta_2 = \theta$. Consequently, the tensions at each end of the rope are also equal: $T_1 = T_2 = T$. Therefore, the relationship between the sideways force and the tension in the rope is given by:

$$F = 2T \sin \theta$$

The angle θ is determined by the geometry of the apparatus. It is given by the expression:

$$\tan \theta = \frac{h}{L}$$

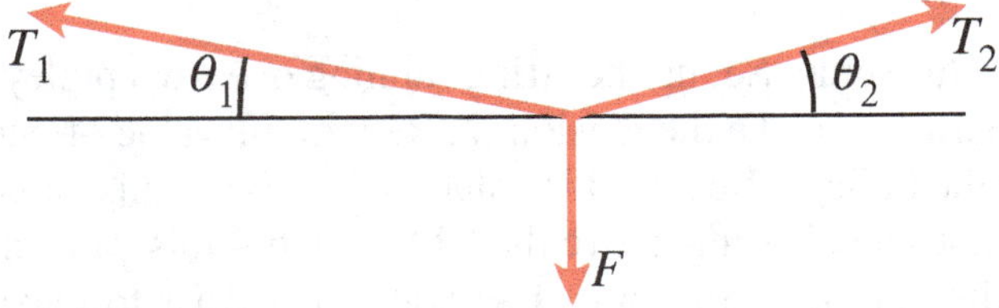

Figure 2.1 Sideways force F causes tension T_1 at deflection angle θ_1 on one end of the clothesline and tension T_2 at deflection angle θ_2 on the opposite end.

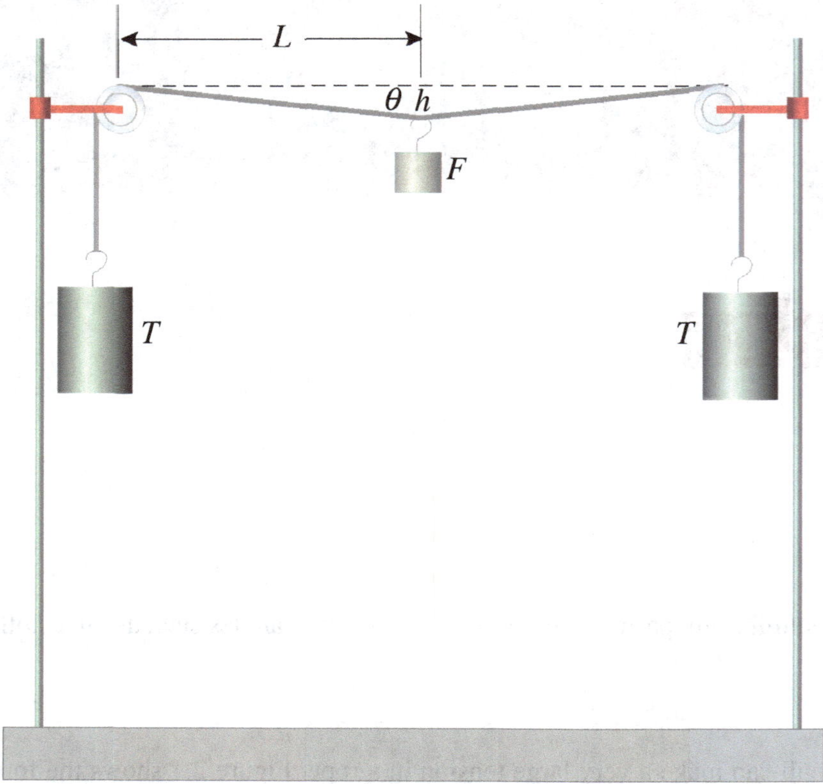

Figure 2.2   Diagram of the laboratory apparatus showing the geometry of the clothesline and the forces acting on it.

where h is the amount of sag in the rope and L is half its length. When performing the error analysis, the standard value is taken as: $F_{std} = 2T \sin \theta$. The corresponding experimental value is determined by the mass attached to the middle of the rope: $F_{exp} = F$. For each trial, the experimental error is given by:

$$\% \text{error} = \frac{F_{exp} - F_{std}}{F_{std}} \times 100$$

Equipment

Assorted masses, string, weight holders, pulleys, and ring stand.

Procedure

1. Figure 2.3 shows the apparatus setup. Set up the string running over two pulleys. Hang two large equal masses on the ends of the string to make the tension, T. Record the value of one of masses as T.
2. Measure the separation of the pulleys. Record this value as $2L$ Divide this value by 2 to determine L.
3. Measure the height of the horizontal string above the table. Record this value as H_1.
4. Hang a weight holder at the middle of the rope. Use additional mass to make-up your desired force increment. Record this value as F.
5. Measure the distance from the table to the middle of the rope. Record this value as H_2.
6. Repeat Steps 4 and 5 four times, increasing the mass in equal increments each time.

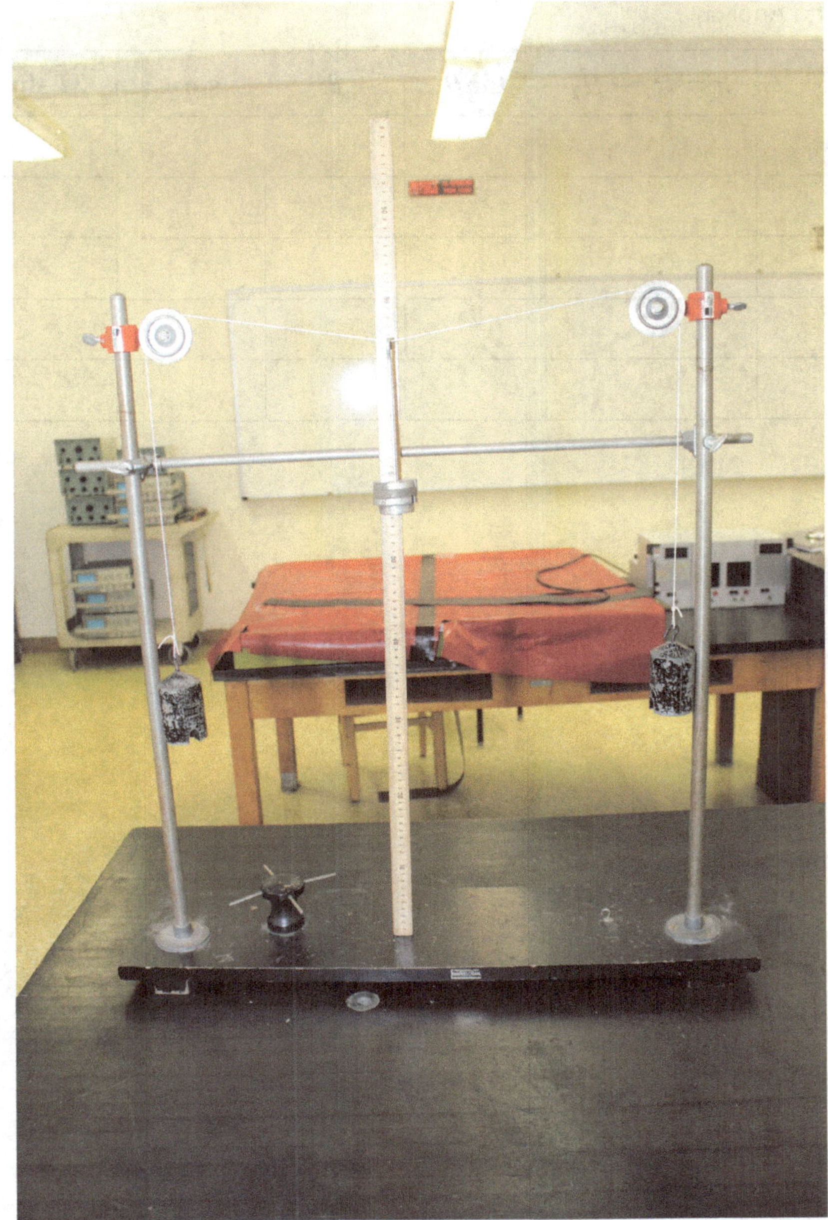

Figure 2.3 The clothesline experimental apparatus has a horizontal string with two equal weights on either end. The weight holder is located at the center of the horizontal segment of the string.

Data Record and Analysis

Record the following values in your laboratory notebook:

$T =$ ________________________ $H_1 =$ ________________________

$2L =$ ________________________ $L =$ ________________________

Reproduce Table 2.1 in our laboratory notebook. Record your data and calculations in their appropriate locations:

Table 2.1   Data and Analysis

F	H_2	$h = H_1 - H_2$	$\tan\theta$	θ	$\sin\theta$	$2T\sin\theta$	% error

Arm Forces

Purpose

To study forces and torques on a model human arm.

Theory

In order for a human to lift an arm, the biceps must exert enough force and torque to overcome the force and torque exerted by the load. Consider the diagram in Figure 3.1. The condition for lifting an arm is given by $F_{\text{lift}} \geq W_{\text{load}}$ and $\tau_{\text{lift}} \geq W_{\text{load}}L$, where F_{lift} is the lifting force, τ_{lift} is the lifting torque, W_{load} is the force exerted by the load, L is the length of the arm.

A diagram of the model arm appears in Figure 3.2. The biceps muscle force is applied close to the pivot. This must lift the load at the end of the arm and the weight of the arm itself. At equilibrium, the torques must balance.

The lifting torque is given by:

$$\tau_{\text{lift}} = m_b a \sin\theta$$

where m_b is the biceps force, a is the distance from the pivot to the biceps attachment point, and θ is the angle between the biceps and the arm. The total load torque is given by:

$$\tau_{\text{load}} = m_a b + m_l c$$

where m_a is the weight of the arm, b is the distance from the pivot to the center of the arm, m_l is the weight of the load, and c is the distance from the pivot to the load.

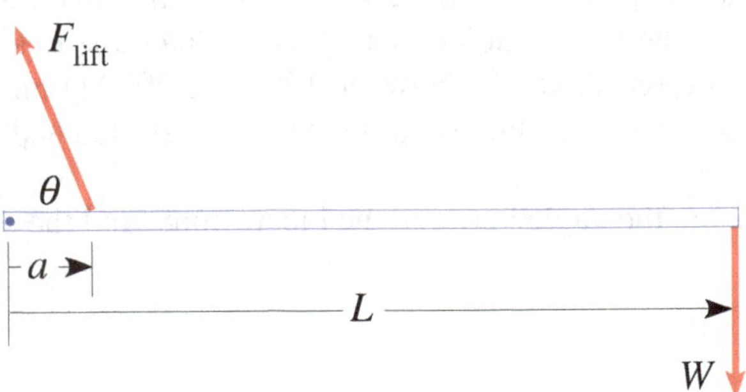

Figure 3.1 Diagram of simplified model of the arm. The weight of the arm is ignored.

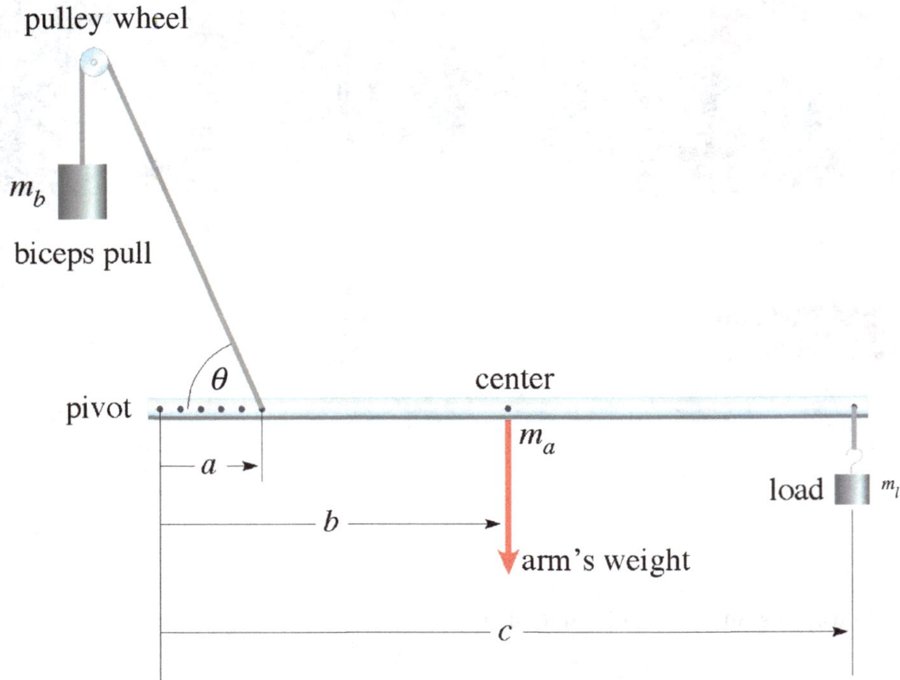

Figure 3.2   A diagram of the arm. The forces acting on the arm, including the load, biceps force, and weight of the arm, are taken into account.

Equipment

Tube, ring stand, pulley, clamps, balance, meter stick, assorted slotted masses, weight holder, and protractor.

Procedure

1. Weigh the forearm tube. Record this value as m_a.
2. Measure the length of the forearm tube. Record this value as L. Divide L by 2 to determine the center of the tube. Mark this spot on the tube.
3. Measure the distance from the pivot to the biceps attachment point. Record this value as a.
4. Measure the distance from the pivot to the center of the tube. Record this value as b.
5. Measure the distance from the pivot to the attachment point of the load. Record this value as c.
6. Assemble the arm. The assembled apparatus is shown in Figure 3.3. The tube used to model the arm should move freely around the pivot with no significant friction. Attach the weight holder to the load attachment point. With only the weight holder acting as the load, the load attachment point should be above the pivot. The biceps force should be between 1000 and 2000 grams. Record this value as m_b.
7. Adjust the load until the arm is horizontal. Record the total mass attached to the load attachment point as m_l.
8. Use the protractor to measure the angle between the biceps force and the arm. Record this value as θ.

Figure 3.3   The model arm force apparatus assembled and balanced.

Data Record and Analysis

Record the following values in your laboratory notebook.

$m_a = $ ____________

$L = $ ____________

$a = $ ____________

$b = $ ____________

$c = $ ____________

$m_b = $ ____________

$m_l = $ ____________

$\theta = $ ____________

$\tau_{\text{lift}} = m_b a \sin\theta = $ ____________

$\tau_{\text{load}} = m_a b + m_l c = $ ____________

The lifting torque is taken as the *experimental value*: $\tau_{\text{exp}} = \tau_{\text{lift}}$. The load torque is taken as the *standard value*: $\tau_{\text{std}} = \tau_{\text{load}}$. The experimental error is taken as:

$$\%error = \frac{\tau_{\text{exp}} - \tau_{\text{std}}}{\tau_{\text{std}}} \times 100\%$$

Range of a Projectile

Purpose

To study projectile motion. Also, to study the dependence of the range on the angle of elevation.

Theory

When an object is projected into the air with a speed v and elevation angle θ, its path follows a curve due to motion in two dimensions. The horizontal (x) component is considered to be at constant or steady velocity. The vertical (y) component is accelerated downward by the constant acceleration of gravity. The trajectory or path of the two combined motions is a parabola. An object following such a path is known as a *projectile*.

When a projectile that has been shot into the air lands on its original level as shown in Figure 4.1, its horizontal displacement is known as its range, R. Under these conditions, the range is given by:

$$R = \frac{v^2 \sin 2\theta}{g}$$

where $g = 9.8 \text{ m/s}^2$ is the acceleration due to gravity. For this laboratory session, this expression will be considered to be $R_{\text{uncorrected}}$, for the uncorrected range.

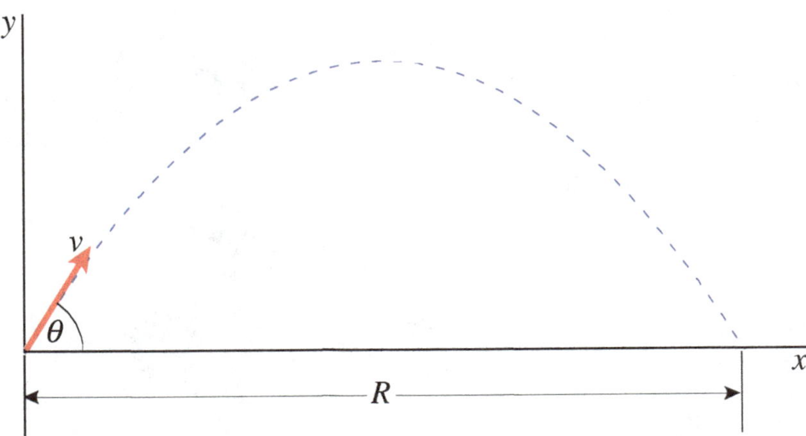

Figure 4.1 The trajectory of a projectile that lands on its projection level after having been projected with velocity v at angle of elevation θ.

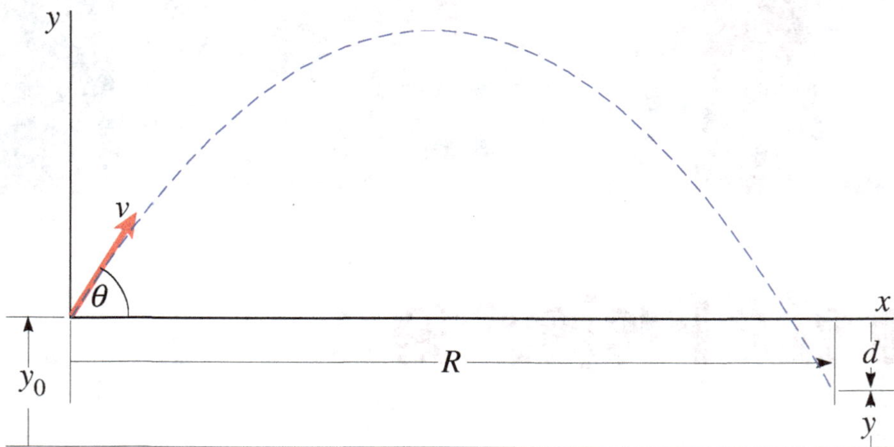

Figure 4.2   The trajectory of a projectile that landed on a level with a vertical displacement d below the projection level having been projected with a velocity v at a projection angle θ.

In the case of a projectile that follows the trajectory diagrammed in Figure 4.2, the projectile lands on a level below the projection level. Its range is given by:

$$R = \frac{v^2 \sin 2\theta}{2g}\left[1 + \sqrt{1 + \frac{2dg}{(v\sin\theta)^2}}\right]$$

where d is the difference between the levels of the projection and landing points. This expression gives $R_{\text{corrected}}$ for the corrected range.

The projection speed v can be determined by shooting the projectile horizontally from a known elevation h and then measuring its horizontal displacement x as shown in the Figure 4.3.

The horizontal displacement x, projection speed v, and time of flight t are related by the equation:

$$x = vt$$

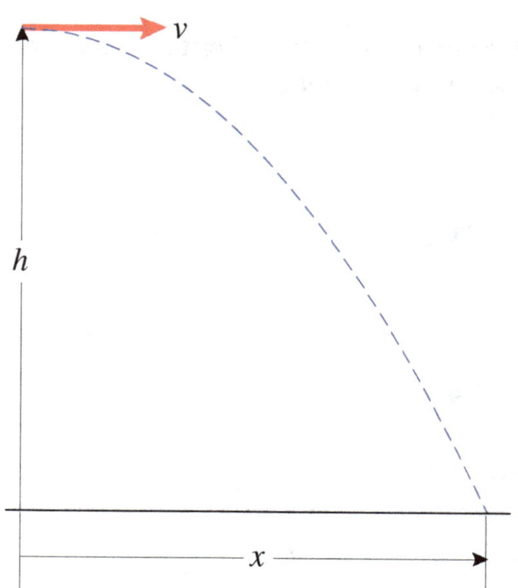

Figure 4.3   The trajectory of a projectile after having been projected with velocity v in the horizontal direction.

The time of flight is determined by the vertical drop h under the influence of gravity:

$$h = \frac{1}{2}gt^2$$

Thus, projection speed is given as:

$$v = x\sqrt{\frac{g}{2h}}$$

Equipment

Spring gun, projectile, tape measure, ruler, riser blocks, adhesive tape, carbon paper, white paper, and inclined plane.

Procedure

1. Locate a safe shooting lane for the projectile to hit the floor.
2. Shoot the projectile once to see where it strikes the floor.
3. Tape a sheet of white paper over this spot. Place a sheet of carbon paper over the white paper.
4. Shoot three additional shots, making sure that the projectile strikes the carbon paper and marks the white paper as shown in Figure 4.4.

Figure 4.4 The projectile strikes carbon paper to mark white paper on the floor to determine the initial velocity of the projectile.

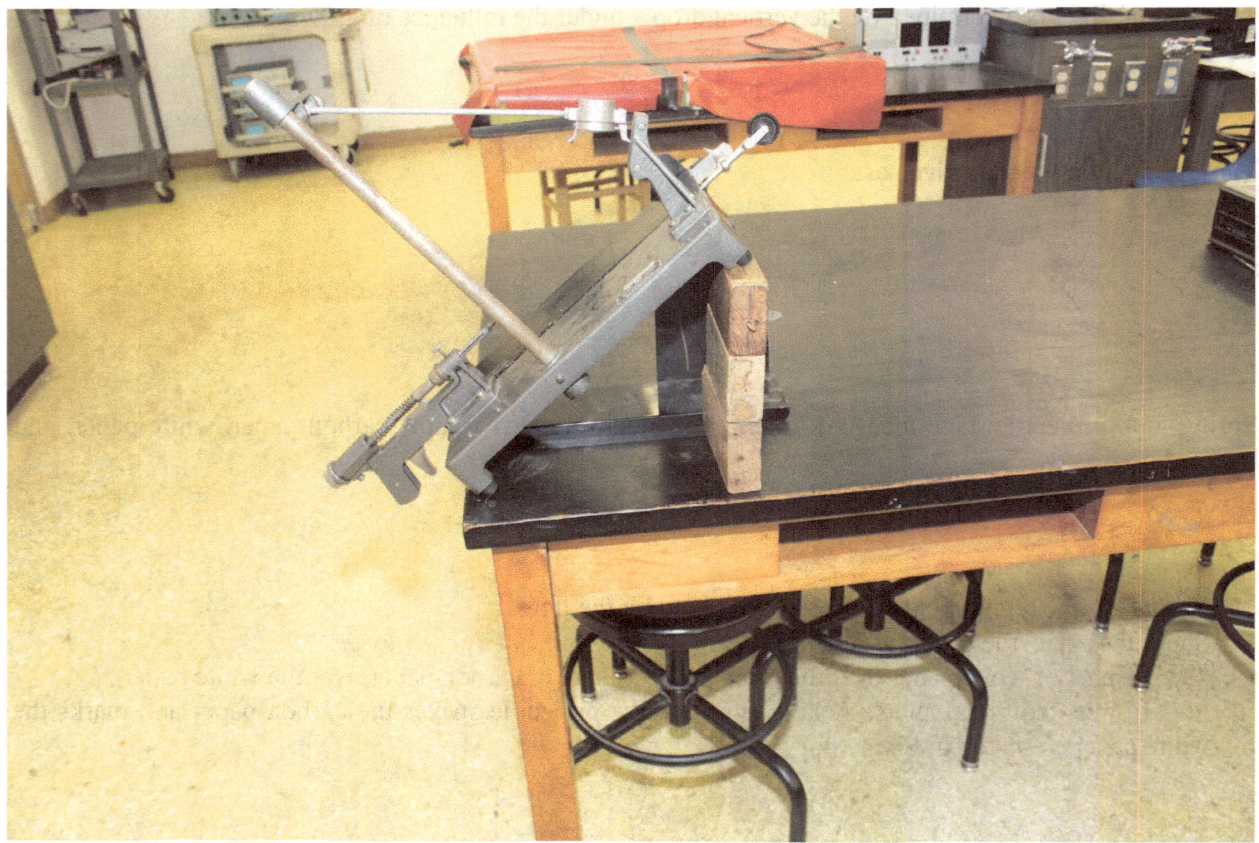

Figure 4.5   Riser blocks determine the elevation angle of the spring gun that is used to launch the projectile.

5. Determine the location on the floor directly beneath the projection point. Measure the elevation from the floor to bottom of the projectile when it is at the projection point. Record this value as h.
6. Use the tape measure to measure the distance from this point to each strike mark. Record these measurements as x_1, x_2, and x_3.
7. Return the gun to your primary workstation.
8. Use a riser block as shown in Figure 4.5 to elevate the front end of the gun.
9. Shoot the projectile to ensure that it strikes the other tabletop in the same row. Tape a sheet of white paper over this spot. Place a sheet of carbon paper over the white paper.
10. Shoot three additional shots, making sure that the projectile strikes the carbon paper and marks the white paper. Figure 4.6 shows the marks on the paper.
11. Measure the elevation between the floor and the bottom of the projectile when it is at the projection point. Record this value as y_0. Measure the elevation between the floor and the landing point. Record this value as y.
12. Determine the point on the tabletop directly beneath the projection point. Use the tape measure to measure the distance from this point to each mark on the white paper. Record these values as R_1, R_2, and R_3.
13. Use the inclined plane to measure the angle of elevation.
14. Repeat Steps 8 through 13 twice, increasing by one the number of riser blocks each time.

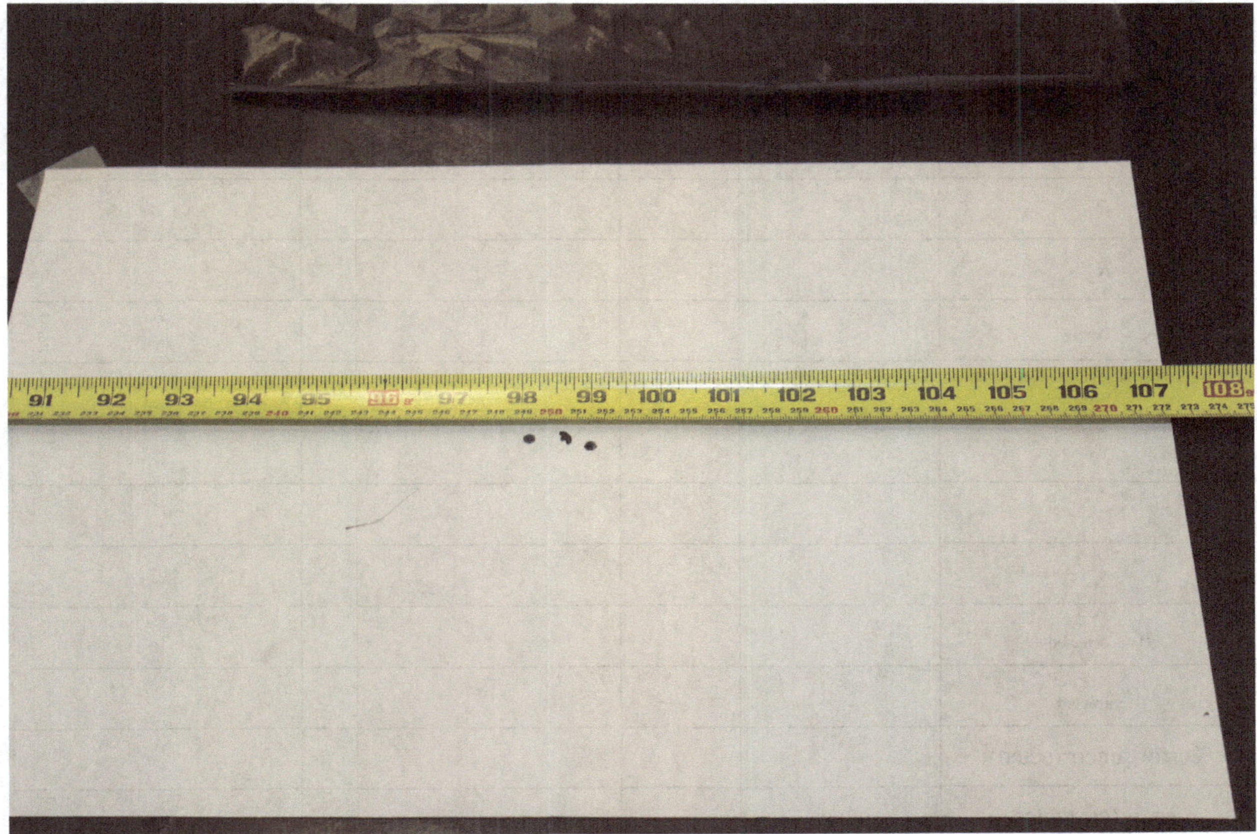

Figure 4.6   The marks created when the projectile strikes carbon paper determine the range of the projectile.

Data Record and Analysis

Record the following values in your laboratory notebook.

$h =$ ____

Reproduce and complete Tables 4.1 and 4.2 in your laboratory notebook.

Table 4.1   Data Required to Determine the Projection Speed of the Projectile

Trial	x_1	x_2	x_3	x_{ave}	$v = x\sqrt{\dfrac{g}{2h}}$

Table 4.2   Projectile Data for Three Angles of Elevation

No. of Riser Blocks	1	2	3
θ			
R_1			
R_2			
R_3			
R_{ave}			
y_0			
y			
$d = y_0 - y$			
$R_{uncorrected}$			
$R_{corrected}$			
%error (uncorrected)			
%error (corrected)			

Friction

Purpose

To study sliding friction and to determine the coefficients of static and kinetic friction.

Theory

Friction is a force that opposes motion. *Static friction* prevents an object's being set into motion by an external force. *Kinetic friction* opposes the progress of a sliding object. Static friction is given by $f_s \leq \mu_s N$, where N is the normal force between the object and its support surface and μ_s is the coefficient of static friction. Therefore, the coefficient of static friction is given by

$$\mu_s = \frac{f_{s,\,max}}{N}$$

where $\mu_{s,\,max}$, the maximum force of static friction. If the external force overcomes static friction, then the object moves. Its motion is opposed by kinetic friction. The force of kinetic friction is given by $f_k = \mu_k N$, where μ_k is the coefficient of kinetic friction. Therefore, the coefficient of kinetic friction is given by:

$$\mu_k = \frac{f_k}{N}$$

A diagram of the experimental setup is shown in Figure 5.1. Of interest is the friction between the block and the tabletop. An external pulling force is provided by the weight hanging on the end of a string that is adjusted to balance the force of friction.

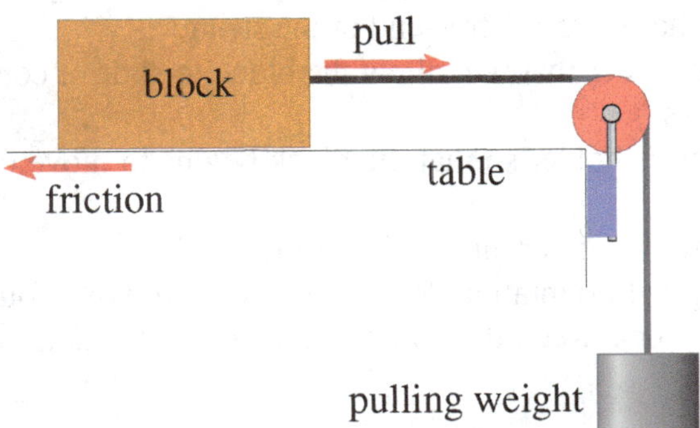

Figure 5.1 A diagram of the friction apparatus. The weight of a hanging mass provides the force that overcomes friction.

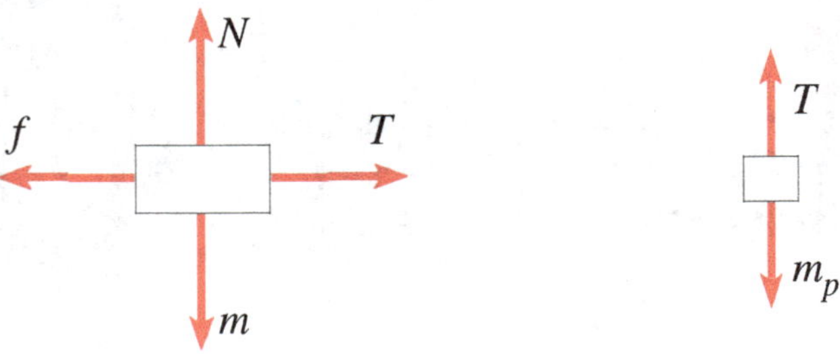

(a) Friction block   **(b)** Pulling weight

Figure 5.2   (a) Free diagram of the forces on the friction block. **(b)** Free diagram of the forces on the pulling weight.

The free-body diagram of the block is shown in Figure 5.2 (a). The free-body diagram of the pulling mass is shown in Figure 5.2 (b). The equations for the forces on the block in the x- and y-directions are written as:

$$\Sigma F_x = 0 = T - f$$
$$\Sigma F_y = 0 = N - m$$

where T is the tension in the string, f is the force of friction, N is the normal force exerted on the block by the tabletop, and m is the total mass supported by the tabletop. The equation for the forces acting on the pulling mass is written as:

$$\Sigma F = 0 = m_p - T$$

Thus, the force of friction f is equal to the pulling mass m_p. The normal force N is given by:

$$N = m = m_b + m_{added}$$

where m_b is the mass of the block and m_{added} is the additional mass that is stacked on the top of the block.

Equipment

Wooden block, pulley, clamps, string, masses, mass holders, and balance.

Procedure

1. Assemble the apparatus as shown in Figure 5.3. Weigh the block. Record this value as m_b.
2. With the block on one of its broad sides, run the string over the pulley. Adjust the string to ensure that it is level. Attach the mass holder to the free end of the string.
3. Find the mass on the mass holder that will allow the block to slide at constant speed once it is set into motion. Record this value as f_k.
4. Add enough mass to the mass holder so that the block begins to move on its own accord. Record the mass as f_s.
5. Repeat Steps 2 through 4 for the other three sides of the block.
6. Return the block to its original orientation. Repeat Steps 2 through 4 four times with an additional 50 to 100 grams added to the block. Record the mass added to the block as m_{added}.

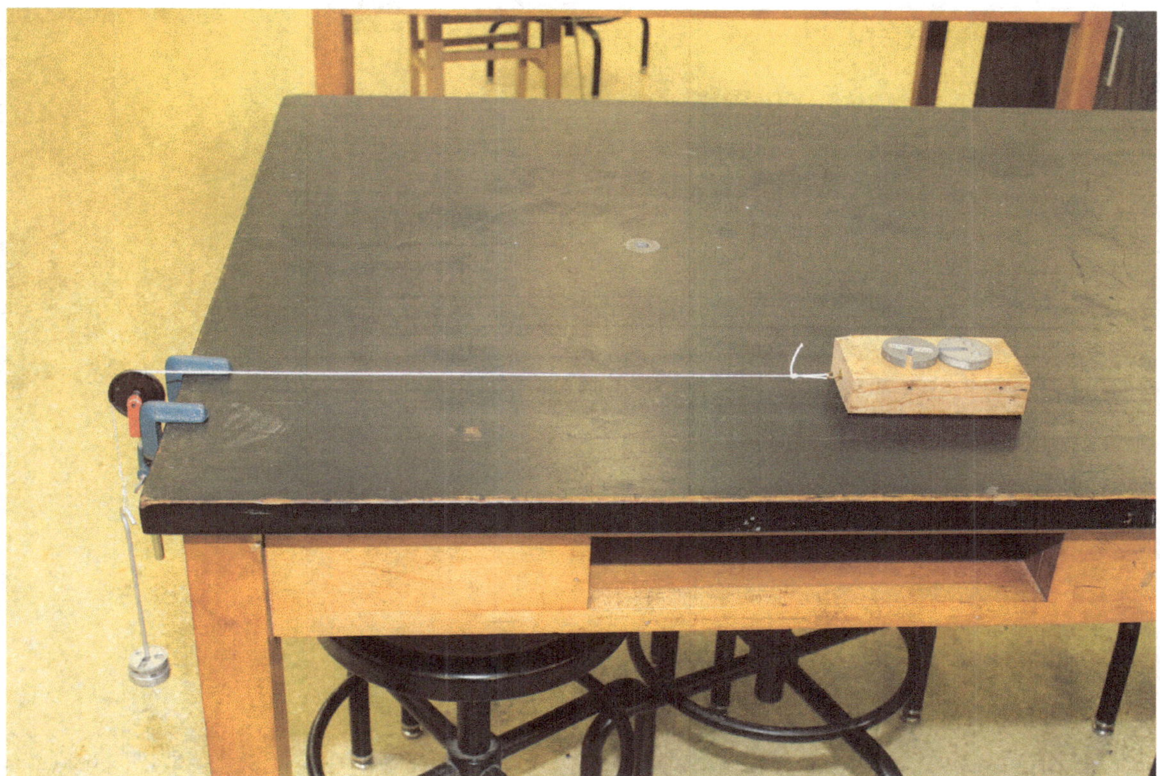

Figure 5.3   A hanging weight pulls a friction block across the laboratory table.

Data Record and Analysis

Record the following values in your laboratory notebook.

mb = _______________________________

Reproduce and complete Table 5.1 in your laboratory notebook.

Table 5.1   Friction Data and Analysis

Side	m_{added}	N	f_k	f_s	μ_k	μ_s
Side 1						
Side 2						
Side 3						
Side 4						
Side 1						
Side 1						
Side 1						
Side 1						
Average (Side 1)						

Centripetal Force

Purpose

To study circular motion and centripetal force.

Theory

If a body is moving with constant speed along a circular path, then it is said to be moving with *uniform circular motion*. Although the speed is constant, the velocity is continually changing, because the direction of the motion is continually changing. Thus, such a body has an acceleration. It can be shown that the direction of the acceleration is always toward the center of the circle, and that the magnitude is given by:

$$a = \frac{v^2}{r}$$

Figure 6.1 The rotating head of the centripetal force apparatus featuring its the mass, pointer, and index to which the pointer points when the rotating mass touches the pointer.

where v is the speed of the body and r is the radius of its circular path. This acceleration may also be expressed in terms of the angular speed, because the linear speed is related to the rotational speed and rotational frequency:

$$v = r\omega$$

where ω is the angular speed in radians per second, and f is the number of revolutions per second. Angular speed ω is related to angular frequency f by $\omega = 2\pi f$ giving:

$$a = \omega^2 r \quad \text{or} \quad a = 4\pi^2 f^2 r$$

A force is necessary to produce this acceleration. This force is a *centripetal force* because it is always directed toward the center of rotation. By Newton's Second Law, $F = ma$:

$$F_c = mr\omega^2 = 4\pi^2 f^2 rm$$

In the apparatus shown in Figure 6.1, the centripetal force of a stretched spring holds the mass in its circular path. When the critical speed is reached, the sensitive pointer will stand opposite the index. This speed is kept constant and is measured by means of a revolution counter and a stop watch.

The force required to stretch the spring by the same amount is subsequently measured by applying weights to the spring. This force is given by $F = (M + m)g$ where M is the additional mass required to stretch the spring and g is the acceleration due to gravity.

Equipment

Centripetal force apparatus, stopwatch, digital caliper, ring stand, mass holder, assorted masses.

Figure 6.2   The centripetal force apparatus with rotating head mounted.

Procedure

1. Turn on the rotator and adjust its rotational speed carefully so that the mass just touches the outer end of the frame and raises the pointer.
2. Record the initial reading of the revolution counter as *Initial Count*. Engage the counter while simultaneously starting the stopwatch. Reset the stopwatch. Count revolutions for approximately 60 seconds.
3. Simultaneously stop the stopwatch and disengage the revolution counter. Record the final reading of the counter as *Final Count*. Record the stopwatch elapsed time as *t*.
4. For each trial, record the difference between *Final Count* and *Initial Count* as Revolutions.
5. Repeat Steps 2 and 3 for two additional trials.
6. Remove the rotating head from the rotating apparatus and hang it from the ring stand as shown in Figure 6.3. Hang mass from the string attached to the rotating mass to stretch the spring just enough to barely touch the pointer. Record this mass as *M*.

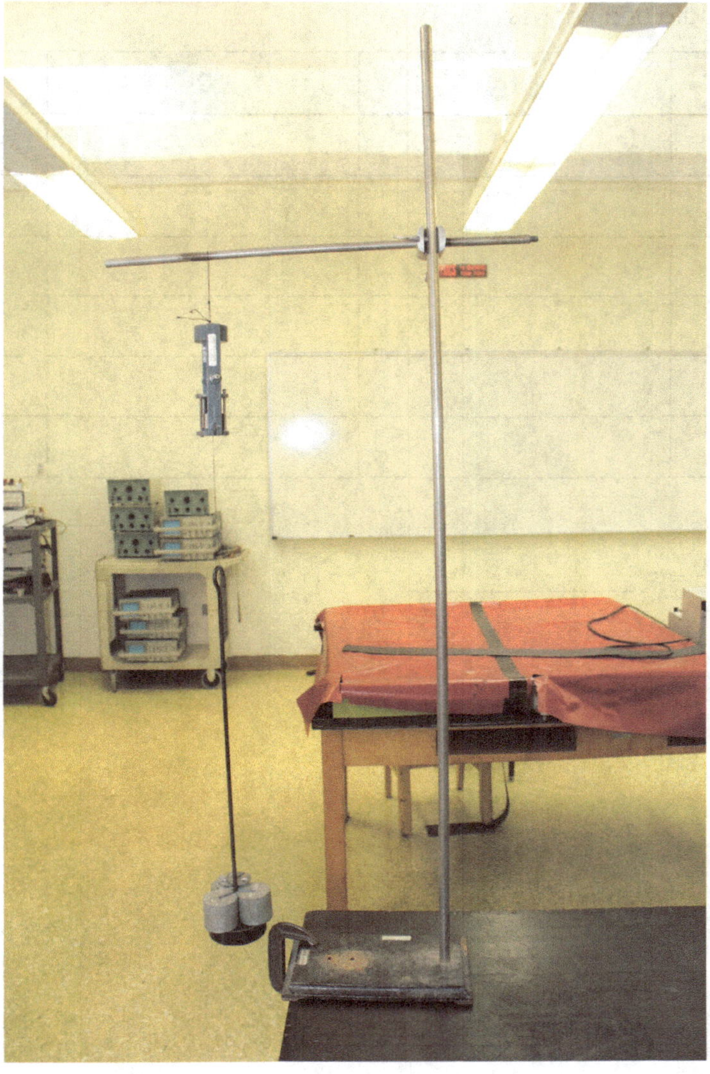

Figure 6.3 The rotating head mounted vertically with sufficient weight applied to fully stretch the spring holding the rotating mass.

7. With the rotating head still suspended, use a digital caliper to measure the distance from the center of the frame to the center of the rotating mass when it just barely touches the pointer. Record this value as r.
8. The value of the small rotating mass is stamped on its end. Locate and record this value as m.

Data Record and Analysis

Record the following values in your laboratory notebook and report:

$M = $ _______________________________

$m = $ _______________________________

$r = $ _______________________________

Reproduce and complete Table 6.1:

Table 6.1   Rotational Data and Analysis

Trial	1	2	3
Initial Count			
Final Count			
Revolutions			
t			
f			
Fc			

$$F_{exp} = \frac{F_{c1} + F_{c2} + F_{c3}}{3} = \underline{\hspace{3cm}}$$

$$F_{std} = (M + m)g = \underline{\hspace{3cm}}$$

$$\%error = \frac{F_{exp} - F_{std}}{F_{std}} \times 100\% = \underline{\hspace{2cm}}$$

Density

Purpose

To study *Archimedes Principle*. To measure the densities of solids.

Theory

Density is defined as the mass per unit volume. In equation form, density is written as:

$$\rho = \frac{m}{V}$$

where ρ is the density, m is the mass, and V is the volume. In the case of a solid, m may be measured using a balance. V may be calculated from the shape and dimensions. For a rectangular parallelepiped, V is the product of the length, width, and height. In equation form, the volume is written as:

$$V = LWH$$

where L is the length, W is the width, and H is the height.

The density may also be determined from an object's specific gravity and the density of water. The specific gravity of an object is the ratio of its density to the density of water:

$$sp.\ gr. = \frac{\rho}{\rho_{water}}$$

where ρ_{water} is the density of water. The value of ρ_{water} is $\rho_{water} = 1.0$ gram/cm^3. Therefore, the density may be determined by the expression:

$$\rho = (sp.\ gr.)\rho_{water}$$

The specific gravity may be determined using *Archimedes Principle*. Archimedes Principle states that an object placed in water loses an amount of weight equal the weight of water that it displaces. The weight of displaced water is the *buoyant force* and is directed upward.

Figure 7.1   A sample is suspended beneath a balance and submerged in a beaker filled with water to determine the buoyant force that water exerts on the sample.

If the sample is suspended beneath the balance and is completely submerged as shown in Figure 7.1, then it experiences three forces: the sample's weight, $m_a g = \rho V g$; the tension in the support string, T; and the buoyant force, $B = m_{\text{water}} g = \rho_{\text{water}} V g$. The tension T is measured using the balance as the sample's weight in water, $m_w g$. The forces are balanced as shown in Figure 7.2:

$$\Sigma F = 0 = T + B - m_a g = m_w g + B - m_a g$$

or

$$B = (m_a - m_w)g$$

The specific gravity may then be written as:

$$sp.\, gr. = \frac{\rho}{\rho_{\text{water}}} = \frac{\rho V g}{\rho_{\text{water}} V g} = \frac{m_a g}{(m_a - m_w)g} = \frac{m_a}{m_a - m_w}$$

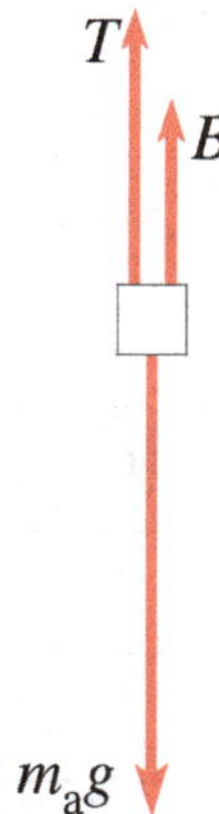

Figure 7.2 The free-body diagram of a sample suspended in water.

Equipment

Metal cubes, balance, digital caliper, ring stand, beaker, water, and assorted masses.

Procedure

1. Identify the composition of your first metal sample. Record it as *Sample Composition*.
2. Use the digital caliper to measure the length, width, and height. Record these values as L, W, and H, respectively.
3. Suspend the sample by a string beneath the balance. Measure its mass. Record this value as m_a.
4. Lift the sample so that a beaker filled with water may fit beneath the balance. Lower the sample into the water while ensuring that it is completely submerged and does not touch the sides of the beaker.
5. Measure the apparent mass of the submerged sample. Record this value as m_w.
6. Repeat Step 1 through Step 5 for your second sample.

Data Record and Analysis

Reproduce and complete Table 7.1:

Table 7.1   Data and Analysis

	1	2
Sample Composition		
ρ_{std} (g/cm^3)		
L (cm)		
W (cm)		
H (cm)		
V (cm^3)		
m_a (g)		
$\rho = \dfrac{m}{V}$		
%error		
m_w (g)		
$sp.\ gr. = \dfrac{m_a}{m_a - m_w}$		
$\rho = (sp.\ gr.)\rho_{water}$		
%error		

Air Track

Purpose

To study head-on collisions and conservation of momentum.

Theory

Momentum is defined as the product of mass and velocity, $p = mv$, where p is the momentum, m is the mass, and v is the velocity. It is a vector quantity that has the same direction as the velocity.

The air track as illustrated in Figure 8.1 provides reduced-friction environment to study head-on collisions. When two or more objects collide, their total momentum is conserved. For example, in this lab, Glider 1 of mass m_1 moving at velocity v_0 will collide with Glider 2, which is at rest. After the collision, Glider 1 moves away with velocity v_1 and Glider 2 moves away with velocity v_2. The total momentum after the collision equals the total momentum before the collision:

$$m_1 v_0 + m_2 0 = m_1 v_1 + m_2 v_2$$

or

$$p_0 = p_1 + p_2$$

where v and p are positive for motion to the right and negative for motion to the left.

Kinetic energy is defined as the energy of motion. For an object of mass m moving with velocity v, its kinetic energy is given by:

$$K = \frac{1}{2}mv^2$$

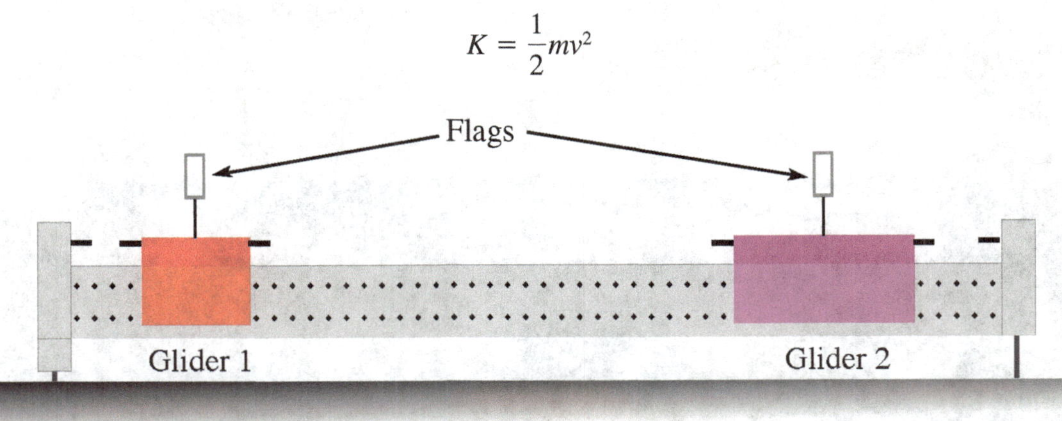

Figure 8.1   The air track provides a nearly friction-free environment to study collisions in one dimension.

In the case of this laboratory experiment, the kinetic energy before the collision is given by:

$$K_0 = \frac{1}{2}m_1 v_0^2$$

The kinetic energy after the collision is given by:

$$K = K_1 + K_2 = \frac{1}{2}m_1 v_1^2 + \frac{1}{2}m_2 v_2^2$$

In the case of an elastic collision, the kinetic energy after the collision (K) equals the kinetic energy before the collision (K_0). In the case of a perfectly inelastic collision, the two gliders stick together after the collision and $v_1 = v_2$. A perfectly inelastic collision results in the maximum reduction of kinetic energy. Most collisions result in a reduction in kinetic energy, but a reduction that is less than maximum.

Equipment

Air track, blower, gliders, balance, Cambridge Physics Outlet (CPO) timers, CPO photogate, assorted masses, and digital caliper.

Procedure

1. Use the balance to measure the masses your gliders. Record their values as m_1 and m_2, respectively.
2. User the digital caliper to measure the width of the flag above each glider. Record their values as Δx_1 and Δx_2, respectively.
3. The assembled air track with gliders is shown in Figure 8.2. Turn on the blower. Set the gliders on the air track. Make sure that each glider moves freely along the track. Make sure that each flag passes through the CPO photogate and activates the photogate.

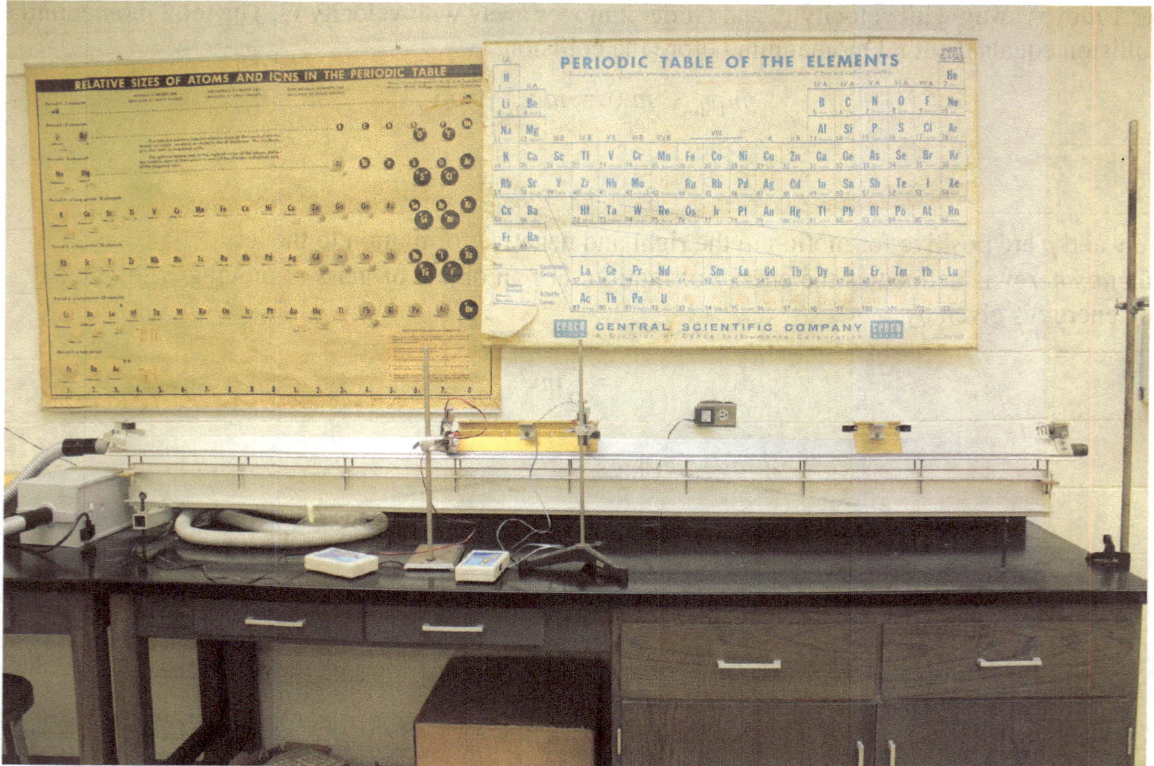

Figure 8.2   Air track with gliders and Cambridge Physics Outlet timers to measure the velocities of the gliders before and after each collision.

4. Set Glider 1 into motion toward Glider 2, which is initially stationary.
5. Use Photogate 1 for Glider 1 before the collision. Use Photogate 2 for Glider 2 after the collision. If Glider 1 moves forward after the collision, then record $(+)$ for After/Glider 1 in the first blank cell of Table 8.1. If it moves backward, then record $(-)$ in the first blank cell.
6. Each photogate records the elapsed times of the last two flag passes. The most recent pass is displayed. The second most recent pass is recorded in memory. If Glider 1 moved forward after the collision, then *Before Glider 1* Δt is displayed by Photogate 1, *After Glider 1* Δt is displayed by Photogate 2, and *After Glider 2* Δt is recorded in Photogate 2's memory. If Glider 1 bounces backward after the collision, then *Before Glider 1* Δt is recorded in Photogate 1's memory, *After Glider 1* Δt is displayed by Photogate 1, and *After Glider 2* Δt is displayed by Photoate 2.

Data Record

Record the following values in your laboratory notebook and laboratory report:

$m_1 = $ _______________________________

$m_2 = $ _______________________________

$\Delta x_1 = $ _______________________________

$\Delta x_2 = $ _______________________________

Table 8.1   Collision Data

	Before	After	
	Glider 1	**Glider 1**	**Glider 2**
$+/-$	$+$		$+$
Δt			
$v = \dfrac{\Delta x}{\Delta t}$			
$p = mv$			
$K = \dfrac{1}{2}mv^2$			

Analysis

Before the Collision

$p_0 = {_1}m_1v_0 = $ _______________________

$K_0 = \dfrac{1}{2}m_1v_0^2 = $ _______________________

After the Collision

$$p = p_1 + p_2 = \underline{\hspace{5cm}}$$

$$K = K_1 + K_2 = \underline{\hspace{5cm}}$$

Error Analysis

$$\%error = \frac{p - p_0}{p_0} \times 100\% = \underline{\hspace{3cm}}$$

$$\%change = \frac{K - K_0}{K_0} \times 100\% = \underline{\hspace{3cm}}$$

Air Table

Purpose

To study the features of a glancing collision in a simulated environment.

Theory

The air table is a two-dimensional version of the air track. The air table allows us to study two-dimensional or *glancing collisions*. Momentum of the discs should be conserved in two dimensions. The total kinetic energy of the two discs may or may not be conserved. A glancing collision is diagrammed in Figure 9.1:

In this collision, Disc 1 has initial velocity v_0. Disc 2 is initially at rest. After the collision, Disc 1 has velocity v_1 with deflection angle θ. Disc 2 has velocity v_2 with deflection angle α.

Momentum is conserved in both the x- and y-direction. In the x-direction, the momentum before the collision equals the momentum after the collision:

$$m_1 v_0 = m_1 v_1 \cos\theta + m_2 v_2 \cos\alpha$$

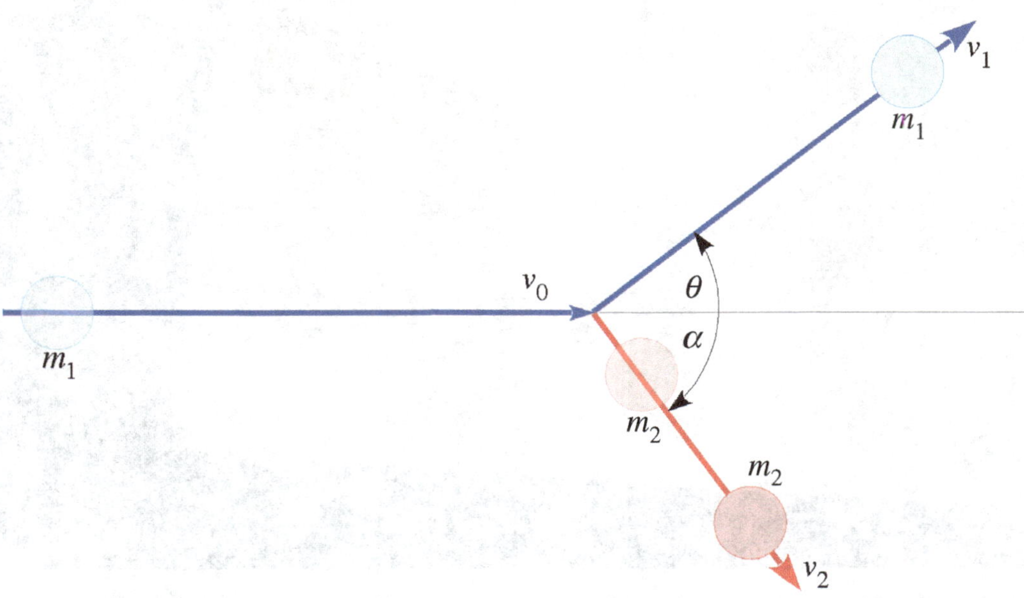

Figure 9.1 A glancing collision in two dimensions.

and in the y-direction, the momentum before the collision equals the momentum after the collision:

$$0 = m_1 v_1 \sin\theta - m_2 v_2 \sin\alpha$$

Kinetic energy. In the case of this laboratory exercise, the kinetic energy before the collision is given by:

$$K_0 = \frac{1}{2} m_1 v_0^2$$

The kinetic energy after the collision is given by:

$$K = K_1 + K_2 = \frac{1}{2} m_1 v_1^2 + \frac{1}{2} m_2 v_2^2$$

In the case of an *elastic collision*, the total kinetic energy after the collision (K) equals the kinetic energy before the collision (K_0). In the case of an elastic collision between discs of equal mass, the discs separate at a right angle to each other.

$$\theta + \alpha = 90°$$

Equipment

Ruler, protractor, simulated time-lapsed photograph.

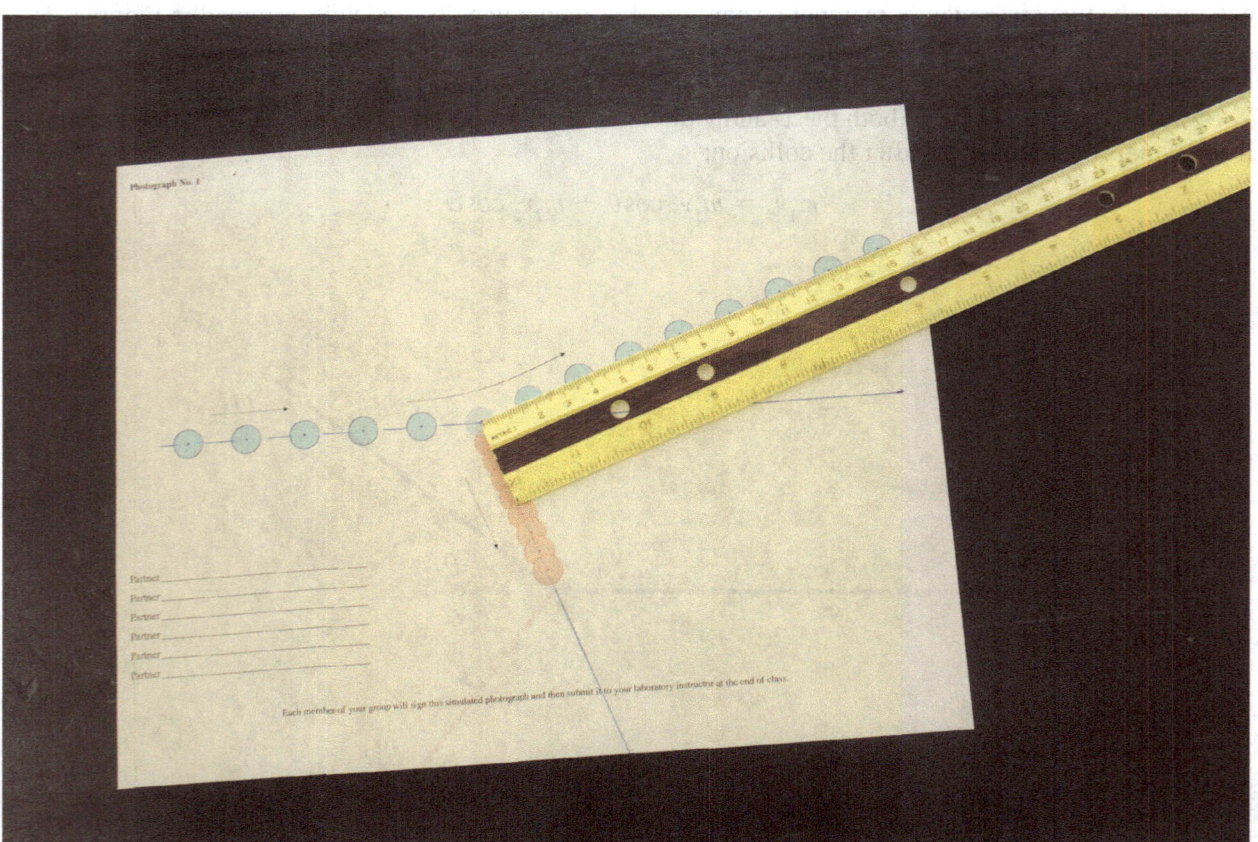

Figure 9.2   A ruler is used to measure the distance between exposures in the simulated time-lapsed photograph.

Procedure

1. Using your simulated time-lapsed photograph and a straight edge, draw a line through the centers of the images of the incident disc. Draw a line through the centers of the images of each disc after the collision.
2. Use the ruler to measure the simulated time-lapsed photograph as shown in Figure 9.2. Measure the distance traveled by the incident disc over five (5) time intervals before the collision. (Each time interval is 0.1 seconds.) Record the distance as d_0 (*Before Disc* 1). Measure the distance traveled by the incident disc over five (5) time intervals after the collision. Record this value as d_1 (*After Disc* 1). Measure the distance traveled by the struck disc over five (5) time intervals after the collision. Record this value as d_2 (*After Disc* 2).
3. Use the protractor to measure the deflection angles of the discs in the simulated time-lapsed photograph as shown in Figure 9.3. Measure θ the angle of deflection of the incident disc. Measure α, the angle of deflection of the struck disc.

Data Record

Record the following values in your laboratory notebook and laboratory report:

$m_1 = 50$ g ________________________________

$m_2 = 50$ g ________________________________

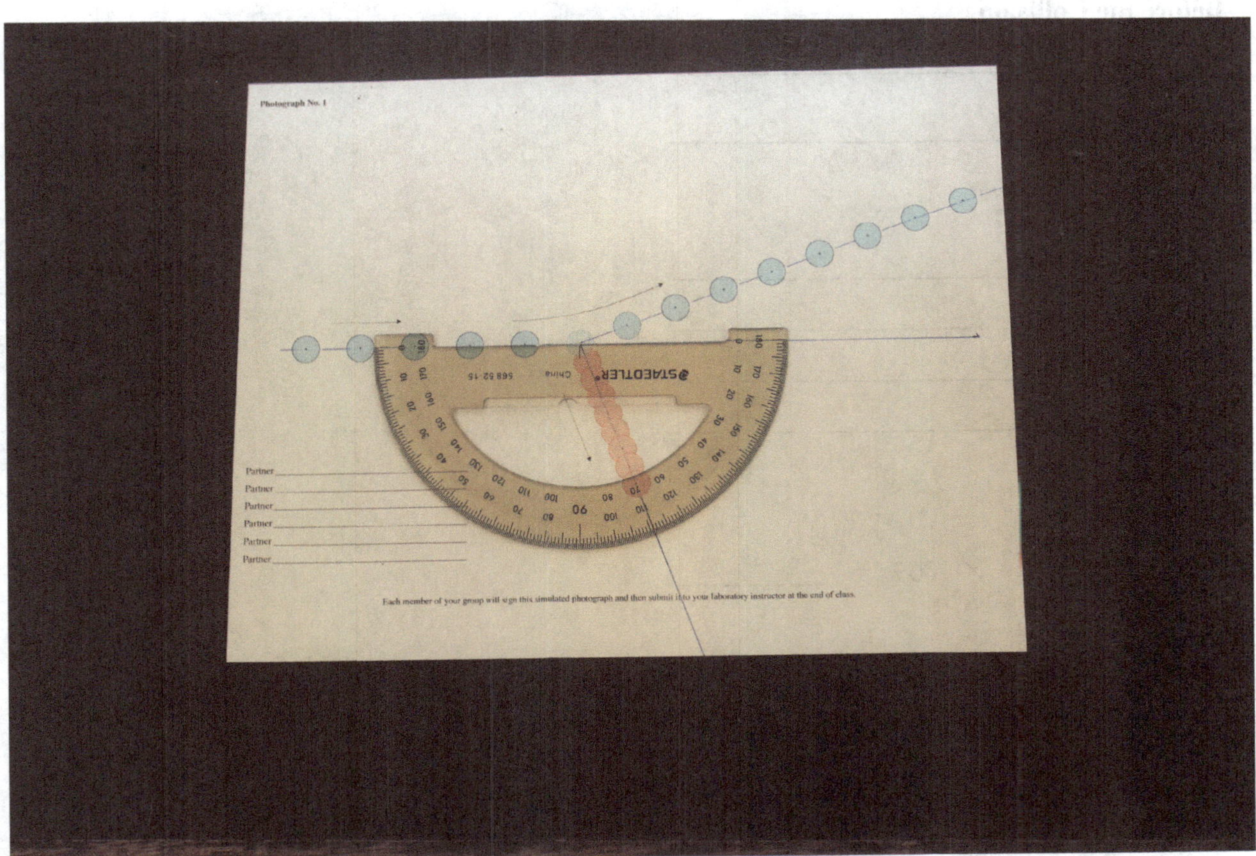

Figure 9.3   The protractor is used to measure the deflection angles of the discs in the simulated time-lapsed photograph.

Table 9.1   Collision Data

	Before	After	
	Disc 1	Disc 1	Disc 2
d			
t	0.5 s	0.5 s	0.5 s
Angle	0.0	$\theta =$	$\alpha =$
$v = \dfrac{d}{t}$			
$p_x = mv_x$			
$p_y = mv_y$	0.0		
$K = \dfrac{1}{2}mv^2$			

Analysis

Before the Collision

$$p_0 = m_1 v_0 \quad = \underline{\hspace{4cm}}$$

$$K_0 = \frac{1}{2}m_1 v_0^2 = \underline{\hspace{4cm}}$$

After the Collision

$$p_x = p_{x1} + p_{x2} = \underline{\hspace{4cm}}$$

$$p_y = p_{y1} + p_{y2} = \underline{\hspace{4cm}}$$

$$K = K_1 + K_2 = \underline{\hspace{4cm}}$$

$$\theta + \alpha = \underline{\hspace{4cm}}$$

Error Analysis

$$\%error_x = \frac{P_x - P_0}{P_0} \times 100\% = \underline{\hspace{3cm}}$$

$$\%error_y = \frac{2p_y}{|p_{y1}| + |p_{y2}|} \times 100\% = \underline{\hspace{2.5cm}}$$

$$\%change = \frac{K - K_0}{K_0} \times 100\% = \underline{\hspace{2.5cm}}$$

$$\%diff = \frac{\theta + \alpha - 90^\circ}{90^\circ} \times 100\% = \underline{\hspace{2cm}}$$

Specific Heat

Purpose

To study heat transfer and to measure the specific heat of a solid.

Theory

When an object changes temperature, it gains an amount of energy equal to $Q = mc(T - T_0)$, where m is the mass of the object, c is its specific heat, T is its final temperature, and T_0 is its initial temperature. The specific heat is of scientific interest because it is a property of the material. Inside a calorimeter, a warm object exchanges thermal energy with a cold object, but there is zero (0) net heat transfer. The process is described by the equation:

$$Q_h + Q_c = 0$$

where Q_h is the energy transferred to the hot object and Q_c is the energy transferred to the cold object. In the case of this experiment, Q_h is the heat transferred to the sample:

$$Q_h = m_h c_h (T - T_h)$$

where m_h is the mass of the sample, c_h is the specific heat of the sample, and T_h is the temperature of the sample. The cold object is a combination of cold water and the aluminum calorimeter cup, which holds the water sample, and stirrer. Therefore, the heat transferred to the cold object is given by:

$$Q_c = (m_c c_{Al} + m_w c_w)(T - T_c)$$

where m_c is the mass of the aluminum calorimeter cup and stirrer, c_{Al} is the specific heat of aluminum, m_w is the mass of the water, and $c_w = 1.0$ cal/g·°C is the specific heat of water. Heat transfer in this experiment is described by:

$$m_h c_h (T - T_h) + (m_c c_{Al} + m_w c_w)(T - T_c) = 0$$

which allows us to calculate the specific heat of the sample:

$$c_h = \frac{(m_c c_{Al} + m_w c_w)\,(T - T_c)}{m_h(T_h - T)}$$

Equipment

Calorimeter, hot plate, metal cubes, water, thermometers, balance, and masses.

Figure 10.1   The balance is used to weigh each sample. The hotplate is used to heat each sample in boiling water to 100 °C. The calorimeter is provides a thermal environment in which heat may transfer among the sample, room temperature water, and aluminum cup with minimal heat transfer with the external environment.

Procedure

1. The equipment used to conduct this laboratory session is shown in Figure 10.1. Begin heating water on the hot plate. Monitor the temperature of the water with a thermometer.
2. Identify the metal used to fabricate your first sample. Record its type as *Composition*. Measure the mass of your sample. Record its value as m_h.
3. Lower the sample into the warming water.
4. Measure the combined mass of the interior calorimeter cup and stirrer. Record its value as m_c.
5. Add water to a depth of approximately 1.0 to 1.5 inches to the calorimeter cup. Measure the combined mass of the calorimeter cup and water. Record this value as $m_c + m_w$. Subtract m_c from this value to determine the mass of the water, m_w.
6. Reassemble the calorimeter. Insert your second thermometer into the cup. Monitor the temperature of the water and calorimeter cup. Record its value as T_c.
7. When the hot water reaches approximately 100 °C, transfer the sample from the hot water to the calorimeter. Record the temperature of the hot water as T_h.
8. Gently stir the water and sample in the calorimeter while you monitor the temperature. Record the peak temperature of the water as T.
9. Disassemble the calorimeter, remove the sample, and dispose of the water.
10. Repeat Step 2 through Step 8 with your second sample.

Data Record and Analysis

Reproduce and complete Table 10.1:

Table 10.1   Calorimetric Data

Sample	1	2
Composition		
c_{std} (cal/g·°C)		
m_h		
m_c		
$m_c + m_w$		
m_w		
T_c		
T_h		
T		
c_h		
$\% error = \dfrac{c_h - c_{std}}{c_{std}} \times 100\%$		

LABORATORY 11

String Waves

Purpose

To study standing waves in a string and to determine their frequency.

Theory

When a wave propagates along a string that is clamped on both ends, it reflects upon itself. As shown in Figure 11.1, standing waves are setup in the string when $n\lambda = 2L$, where $n = 1, 2, 3, 4, \ldots$ is the number of antinodes, λ is the wavelength, and L is the distance between the string supports. The string vibrates in a series of *nodes* and *antinodes* as shown in Figure 11.2. Nodes are points that remain fixed. The attachment points are nodes. Antinodes are points on the string that vibrate with maximum amplitude.

The velocity of the wave in the string is given by:

$$v = \sqrt{\frac{T}{\mu}}$$

where T is the tension in the string and μ is its linear mass density. The tension is the weight of the total mass hanging over the pulley, $T = Mg$. The string's linear mass density is given by $\mu = m/\ell$. The frequency, velocity, and wavelength are related to each other by the expression: $f = v/\lambda$.

Equipment

Electric vibrator, string, clamps, weight holder, pulley, assorted masses, analytic balance, and tape measure.

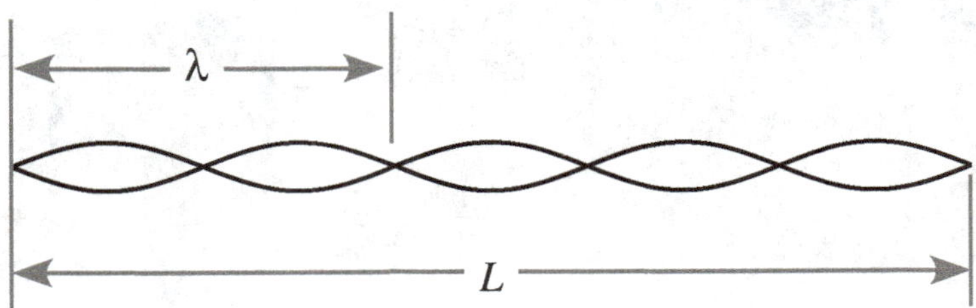

Figure 11.1　The length of a vibrating string at resonance is equal to an integral number of half-wavelengths.

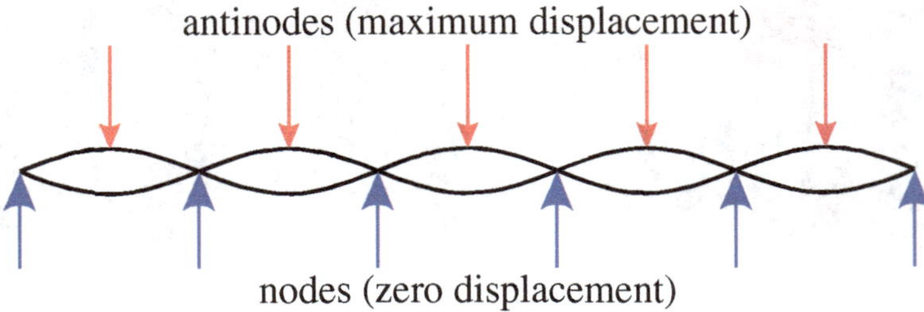

Figure 11.2   The standing wave structure in a vibrating string at resonance.

Procedure

1. Assemble the apparatus. Clamp the electric vibrator to one end of the lab table. Clamp the pulley directly across from the vibrator on the other end of the table. Tie a string to the vibrator and run it over the pulley. Attach the weight holder to the end hanging over the pulley.
2. Plug in the vibrator.
3. Add masses to the weight holder until the vibrating segments are clearly visible. You should see several such segments with such segments when only the mass holder is attached to the free end of the string. Adjust the mass on the weight hanger until the amplitude of the vibrating segments is maximum. Figure 11.3 shows the string vibrating at resonance.
4. Record the number of vibrating segments as n and the mass including the weight holder as M.
5. Repeat Steps 3 and 4 to find a lesser number of vibrating antinodes.

Figure 11.3   A string vibrating at resonance.

6. Measure the length of the string from the string attachment point on the vibrator to the top of the pulley. Record this value as L.
7. Cut the string from the vibrator and the weight holder. Measure the length of loose string. Record this value as ℓ.
8. Use the analytic balance to measure of the mass of the string. Record this value as m.

Data Record and Analysis

Record the following data and calculations in your laboratory notebook:

$L =$ _______________________________

$m =$ _______________________________

$\ell =$ _______________________________

$\mu = \dfrac{m}{e} =$ _______________________________

Reproduce Table 11.1 in your laboratory notebook. Record your data and calculations in their appropriate locations:

Calculate the average frequency. In your notebook, record this as your *experimental value*:

$f_{exp} =$ _______________________________

Your standard value may be taken as: $f_{exp} = 120.0$ Hz.

Your experimental error is calculated as follows:

$$\%\,error = \frac{f_{exp} - f_{std}}{f_{std}} \times 100\% = \underline{\qquad}$$

Table 11.1:   Vibration Data

n	M	$T = Mg$	$\lambda = \dfrac{2L}{n}$	$v = \sqrt{\dfrac{T}{\mu}}$	$f = \dfrac{v}{\lambda}$

Resonance

Purpose

To study standing sound waves in a closed-in tube and resonance in sound waves.

Theory

Sound waves in a closed-end tube reflect upon themselves. Under certain conditions, the waves form standing waves and resonate in the tube. Standing waves in a closed-end tube are illustrated in Figure 12.1. The closed end of the tube must be a *node* where the oscillation is minimum. The open end of the tube must be an *antinode* where the vibration is maximum. The condition for standing waves or resonance is given by:

$$L = \frac{m\lambda}{4}$$

where L is the length of the column of air in the tube, λ is the wavelength, and $m = 1, 3, 5, \ldots$.

The wavelength of the wave, the wave speed, and the wave frequency are related by the expression $v = f\lambda$, where v is the speed of sound, f is the frequency, and λ is the wavelength. The room temperature speed of

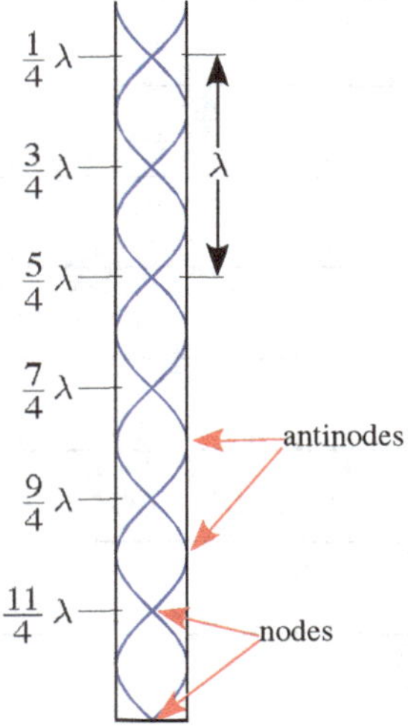

Figure 12.1 The mode structure of sound waves at resonance in a closed end tube.

sound is approximately $v = 343$ m/s. The frequency is determined by the setting of the audio oscillator. The wavelength is twice the distance between adjacent nodes.

Equipment

Resonance tube, audio oscillator, speaker, and water.

Procedure

1. Fill the resonance tube so that the water level is near the top of the tube and the bottom of the reservoir.
2. Plug the speaker into the sine wave output of the audio oscillator.
3. Set the oscillator in the 10 V range.
4. Plug the oscillator into the electrical outlet and raise the amplitude until the speaker is audible.
5. Hold the speaker over the open end of the tube.
6. Slowly move the reservoir down until the sound reaches maximum intensity. Adjust the reservoir for maximum sound intensity. Record the position of the maximum as L.
7. Repeat Step 6 for the remaining resonances.

Data Record and Analysis

Record the frequency setting of your audio oscillator as your *standard value*

$f_{std} = $ ______________________________________

Reproduce Table 12.1 in your laboratory notebook. Record your data and calculations in their appropriate locations:

Table 12.1   Resonance Data

M	L	ΔL
1		
3		
5		
7		
9		
11		
13		

Figure 12.2   Assembled resonance tube partially filled with water. The water level determines which frequencies of sound will produce standing waves above the water in the tube.

Calculate the average value of ΔL. Record this value and the other values as indicated in your laboratory notebook:

$$\Delta L_{ave} = \underline{\hspace{5cm}}$$

$$\lambda = 2\Delta L_{ave} = \underline{\hspace{5cm}}$$

Calculate and record the experimental value of the frequency of your sound waves:

$$f_{exp} = \frac{v}{\lambda} = \underline{\hspace{5cm}}$$

Your experimental error is calculated as follows:

$$\%\,error = \frac{f_{exp} - f_{std}}{f_{std}} \times 100\% = \underline{\hspace{5cm}}$$

Ohm's Law

Purpose

To study how well Ohm's Law predicts the resistance of various resistors and resistor combinations. Also, to learn how to operate a DC power supply and digital multimeter and to read the carbon resistor color code.

Theory

A large class of materials allows the movement of charge or electric current when a difference in potential is applied across them. This phenomenon is described by Ohm's Law:

$$V = IR$$

where V is the difference in potential in volts (V), I is the electric current in amperes (A), and R is the resistance in ohms Ω. Devices known as *resistors* are commonly used as components in electric and electronic circuits.

Resistors may be combined to produce resistance values other than those of the available discrete components. The two simplest combinations are the *series* combination and the *parallel* combination. The combinations are diagrammed in Figure 13.1. In the case of two resistances, R_1 and R_2, in series combination, the equivalent resistance is given by:

$$R_s = R_1 + R_2$$

In the case of the same two resistances in parallel combination, the equivalent resistance is given by:

$$R_p = \frac{R_1 R_2}{R_1 + R_2}$$

Resistors are usually made of carbon or of coils of finely wound wire. Wire-wound resistors are usually reserved for high-precision or high-power applications. Carbon resistors are used in most applications.

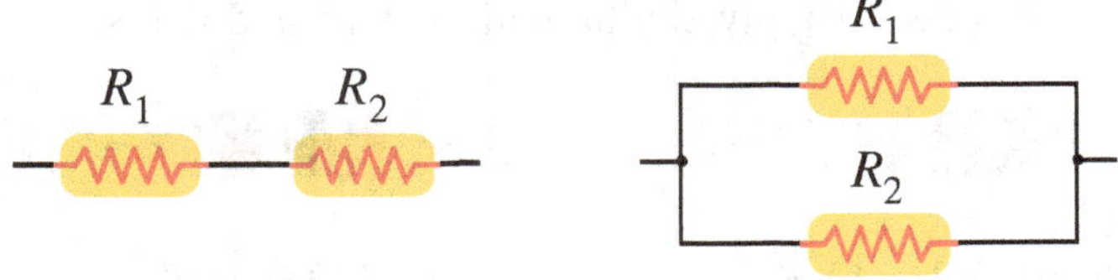

(a) Series Combination   (b) Parallel Combination

Figure 13.1 Simple combinations of resistors R_1 and R_2.

Table 13.1   Color Code Numerical Values

| Color | First Three Bands | | | Fourth Band |
	1ˢᵗ Digit	2ⁿᵈ Digit	Multiplier	Tolerance
Black	0	0	10^0	$\pm 1.0\%$
Brown	1	1	10^1	$\pm 2.0\%$
Red	2	2	10^2	
Orange	3	3	10^3	
Yellow	4	4	10^4	
Green	5	5	10^5	$\pm 0.5\%$
Blue	6	6	10^6	$\pm 0.25\%$
Violet	7	7	10^7	$\pm 0.10\%$
Gray	8	8	10^8	$\pm 0.05\%$
White	9	9	10^9	
Gold			10^{-1}	$\pm 5.0\%$
Silver			10^{-2}	$\pm 10.0\%$
Colorless				$\pm 20.0\%$

Carbon resistors are commonly marked with three or four color bands. The color of each band and its placement indicate the value of the resistance according to the color code. The color code is listed in Table 13.1.

Consider Example (a) in Figure 13.2. The first band is blue which gives "6" as the first digit. The second band is gray, which gives "8" as the second digit. The third band is orange which means the multiplier is 10^3. The fourth band is gold which means that the color code is accurate to within 5%. Therefore, this resistor has a resistance of $68 \times 10^3 \Omega \pm 5\%$ or $68\ \text{k}\Omega \pm 3\ \text{k}\Omega$. Now consider Example (b) in Figure 13.2. Its first band is red which gives "2" as the first digit. The second digit is "7" because the band is violet. The multiplier is 10^0 or "1" because the third band is black. The fourth band is silver, giving a tolerance of 10%. Therefore, this resistor has a resistance of $27\ \Omega \pm 10\%$ or $27\ \Omega \pm 3\ \Omega$.

Equipment

Carbon resistors, breadboard, DC power supply, digital multimeters, and wires.

(a) $68 \times 10^3\ \Omega \pm 5\%$ **(b)** $27\ \Omega \pm 10\%$

Figure 13.2   Two examples of common carbon resistors, each with four color bands that represent their resistance values and tolerances.

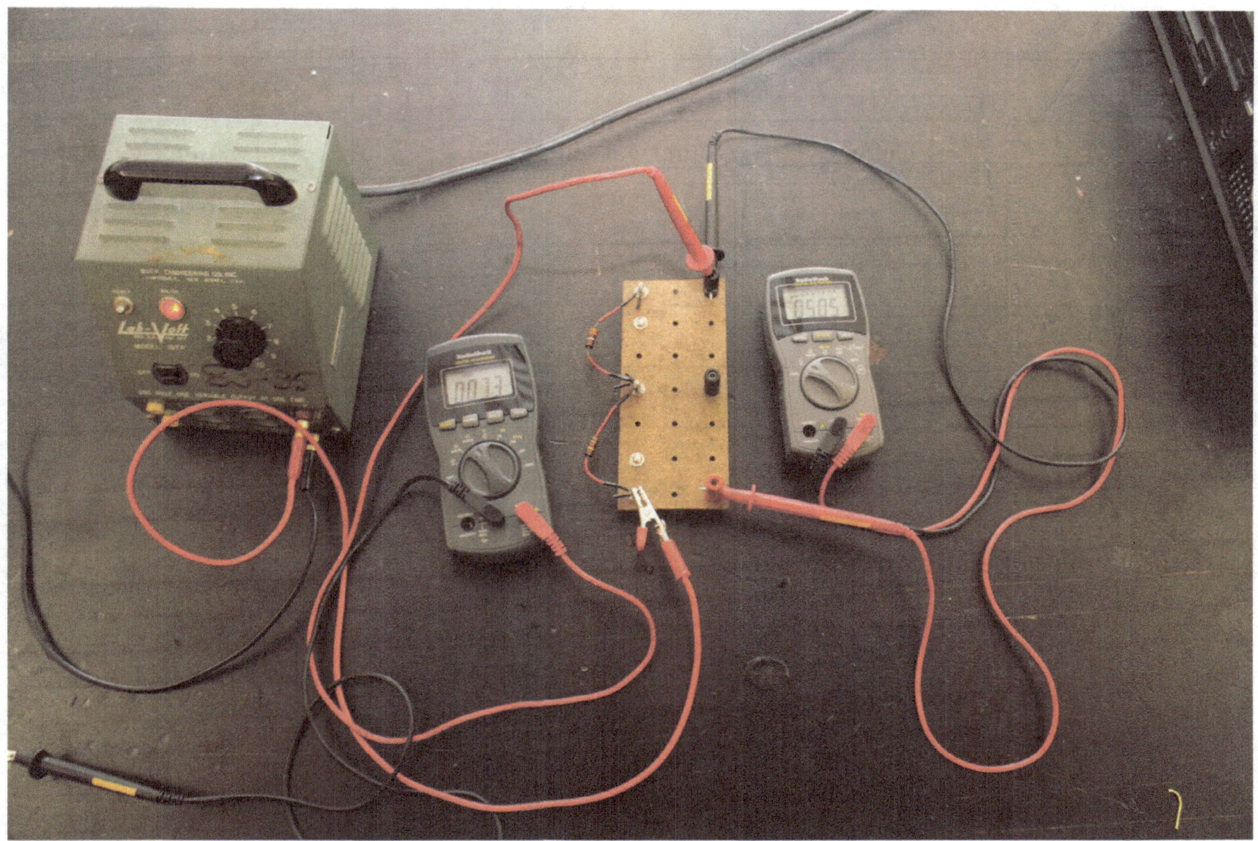

Figure 13.3   The current through two and the potential difference across two resistors connected in series are measured simultaneously.

Procedure

1. Turn on the DC power supply. Connect the digital multimedia to the power supply. Adjust the power supply to an output of 5 V.
2. Setup the first resistor on the breadboard in series with the power supply. Record its color code.
3. Connect the multimeter in parallel with the resistor. Measure and record the voltage across the resistor.
4. Rearrange the connection so that the multimeter is in series with the resistor and power supply.
5. Measure and record the current through the resistor.
6. Repeat Steps 2 through 5 with the second resistor.
7. Setup the two resistors in series on the breadboard.
8. Repeat Steps 2 through 5 with the combination of resistors.
9. Setup the two resistors in parallel on the breadboard.
10. Repeat Steps 2 through 5 with the combination.

Data Record and Analysis

Reproduce Table 13.2 in your laboratory notebook. Record the color codes of your two resistors:

Table 13.2   Resistor Color Codes

Band	R_1		R_2	
	Color	**Numerical Value**	**Color**	**Numerical Value**
First				
Second				
Third				
Fourth				
Numerical Value				

The standard value of each resistor is the numerical value that is determined by its color code:

$R_{1,\text{std}} = $ ______________________________

$R_{2,\text{std}} = $ ______________________________

The standard values of the resistors in combination are calculated from these values:

$R_{s,\text{std}} = R_1 + R_2 = $ ______________

$R_{p,\text{std}} = \dfrac{R_1 R_2}{R_1 + R_2} = $ ______________

Reproduce Table 13.3 in your laboratory notebook. Record your data and results in the places indicated:

Table 13.3   Measured Values

Resistor	V	I	$R_{\text{exp}} = \dfrac{V}{I}$	%error
R_1				
R_2				
R_s (series)				
R_p (parallel)				

Wheatstone Bridge

Purpose

To study the Wheatstone bridge circuit. Also, to use the Wheatstone bridge to measure the resistance of conductors and to determine their resistivities.

Theory

A Wheatstone bridge is a widely used circuit for measuring resistance accurately. The circuit is diagrammed in Figure 14.1. It consists of four resistors, three of which are known, and a fourth which is to be measured. The four resistors are connected into a bridge, $ABCD$, as shown in the diagram. A power supply is connected across points A and C. A galvanometer is connected across points B and D. A galvanometer is a very sensitive instrument which indicates the flow of current. The resistances are adjusted so that the current through the galvanometer is zero. Nothing crosses over the bridge. In this condition, the bridge is said to be balanced and the value of the unknown resistance can be determined.

When the bridge is balanced, I_1, the current through R_1, is also the current through R_2. The current I_2, the current through R_3, is also the current through R_4. Points B and D are at the same potential. This gives rise to two equations:

potential difference between A and B (V_1) = potential difference between A and D (V_3)

$$I_1 R_1 = I_2 R_3$$

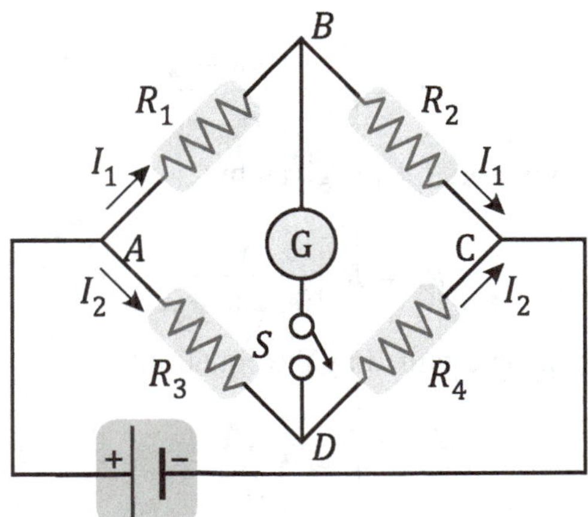

Figure 14.1 The diagram of a Wheatstone bridge connected to a battery.

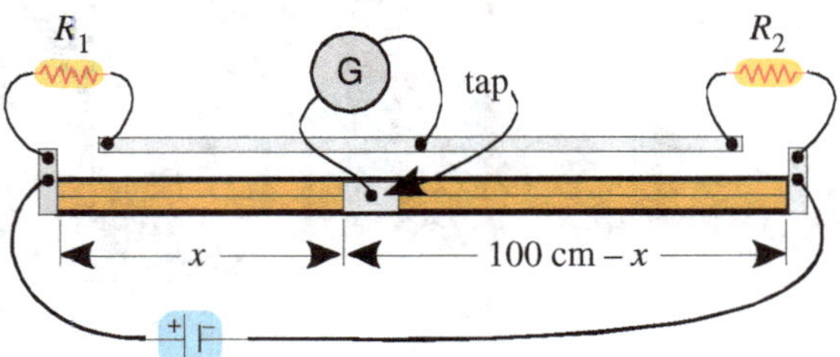

Figure 14.2   Diagram of slidewire Wheatstone with sample resistance R_1 and reference resistance R_2.

potential difference between B and C (V_2) = potential difference between D and C (V_4)

$$I_1 R_2 = I_2 R_4$$

Divide the first equation by the second to give:

$$\frac{R_1}{R_2} = \frac{R_3}{R_4}$$

which gives R_1 as:

$$R_1 = R_2 \frac{R_3}{R_4}$$

This laboratory is conducted using a slidewire Wheatstone bridge as shown in the diagram in Figure 14.2. R_2 is a fixed resistance. R_3 is the resistance of the slidewire between its positive end and the tap. This resistance is given by:

$$R_3 = \rho \frac{x}{A}$$

where ρ is the resistivity of the slidewire, x is the distance between its positive end and the tap, and A is its cross-sectional area. R_4 is the resistance of the slidewire between its negative end and the tap. This resistance is given by:

$$R_4 = \rho \frac{100\ \text{cm} - x}{A}$$

R_1 is the unknown resistance to be measured. R_1 is given by:

$$R_1 = R_2 \frac{x}{100\ \text{cm} - x}$$

The resistivity of the spool of wire is determined by the equation:

$$\rho = R \frac{A}{L}$$

where R is the resistance as measured using the Wheatstone bridge, A is the cross-section area as determined from the labeled gauge of the wire, and L is the labeled length of the wire.

Figure 14.3 A slidewire Wheatstone bridge is assembled and ready to measure the resistance of resistance spools.

Equipment

Slidewire Wheatstone bridge, power supply, galvanometer, resistance spools, high-precision resistor, and assorted patch wires.

Procedure

1. Record the value of the fixed resistor as R_2.
2. Assemble the bridge circuit. Connect Spool 1 to the circuit. Record its conductor, length L, and diameter d. Figure 14.3 shows the slidewire Wheatstone bridge assembled with resistance spools, power supply, and galvanometer.
3. Adjust the position of the tap until the galvanometer needle shows no deflection.
4. Record the position of the tap as x.
5. Repeat Steps 2 through 4 for each of the remaining four spools.

Data Record and Analysis

Record the value of the fixed resistor in your laboratory notebook:

$R_2 = $ _______________________________________

Reproduce Table 14.1 in your laboratory notebook:

Table 14.1   Wheatstone Bridge Data

Spool	Conductor	L	d	x	R	ρ	%error
1							
2							
3							
4							
5							

LABORATORY 15

Oscilloscope

Purpose

To study the features and operation of oscilloscopes.

Theory

Figure 15.1 features a typical cathode ray oscilloscope. An oscilloscope displays an image of electrical effects. It does this so quickly that it can reveal the detailed variations in rapidly changing electrical signals. Therefore, the oscilloscope has many practical applications and is useful in any laboratory or electronics repair shop.

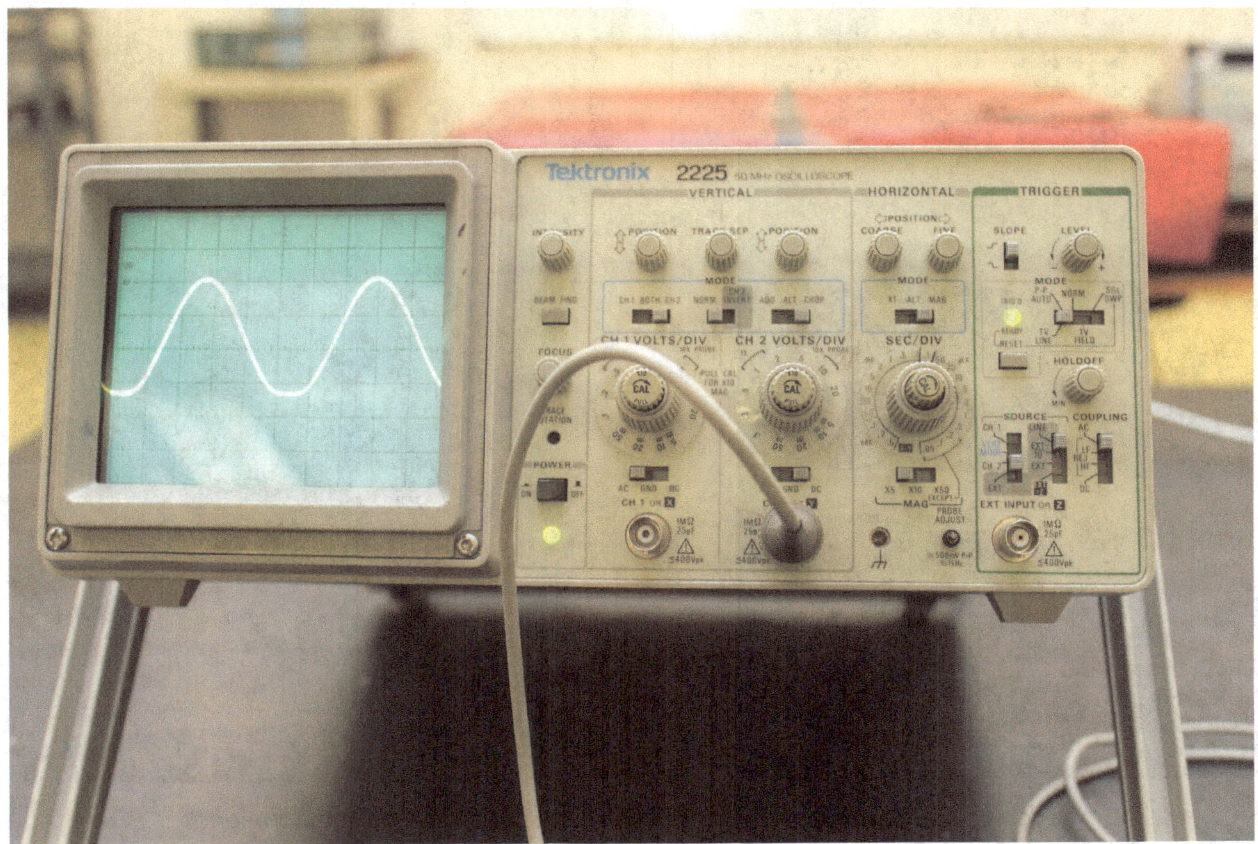

Figure 15.1   An oscilloscope displays a sine wave signal on Channel 2.

The oscilloscope shoots a beam of electrons toward the screen in the front where they make a bright green (or blue) spot. On its way toward the screen, the beam passes through two sets of deflecting plates. The horizontal deflecting plates sweep the beam gradually from left to right and then send it back instantly to the left side of the screen. You control the rate with a SEC/DIV dial, usually on the right side of the oscilloscope.

The vertical deflecting plates receive the voltage signal you want to study. Your voltage moves the beam up and down while it steadily sweeps across the screen. You may control the amplitude of the up and down motion with a VOLTS/DIV dial on the left side of the oscilloscope.

Equipment

Tektronix oscilloscope, audio oscillator, coaxial probe, and wires.

Procedure

1. Turn on the oscilloscope and the audio oscillator. Allow them both to warm up. Figure 15.2 displays the oscilloscope receiving a signal from the sine wave output terminal of an audio oscillator.
2. Adjust the FOCUS and INTENSITY so that the trace is clearly visible, but not too intense.
3. Input a sine wave signal to the oscilloscope.
4. Adjust the VOLTS/DIV knob. What happened?
5. Adjust the SEC/DIV knob. What happened?
6. Adjust the vertical position. What happened?

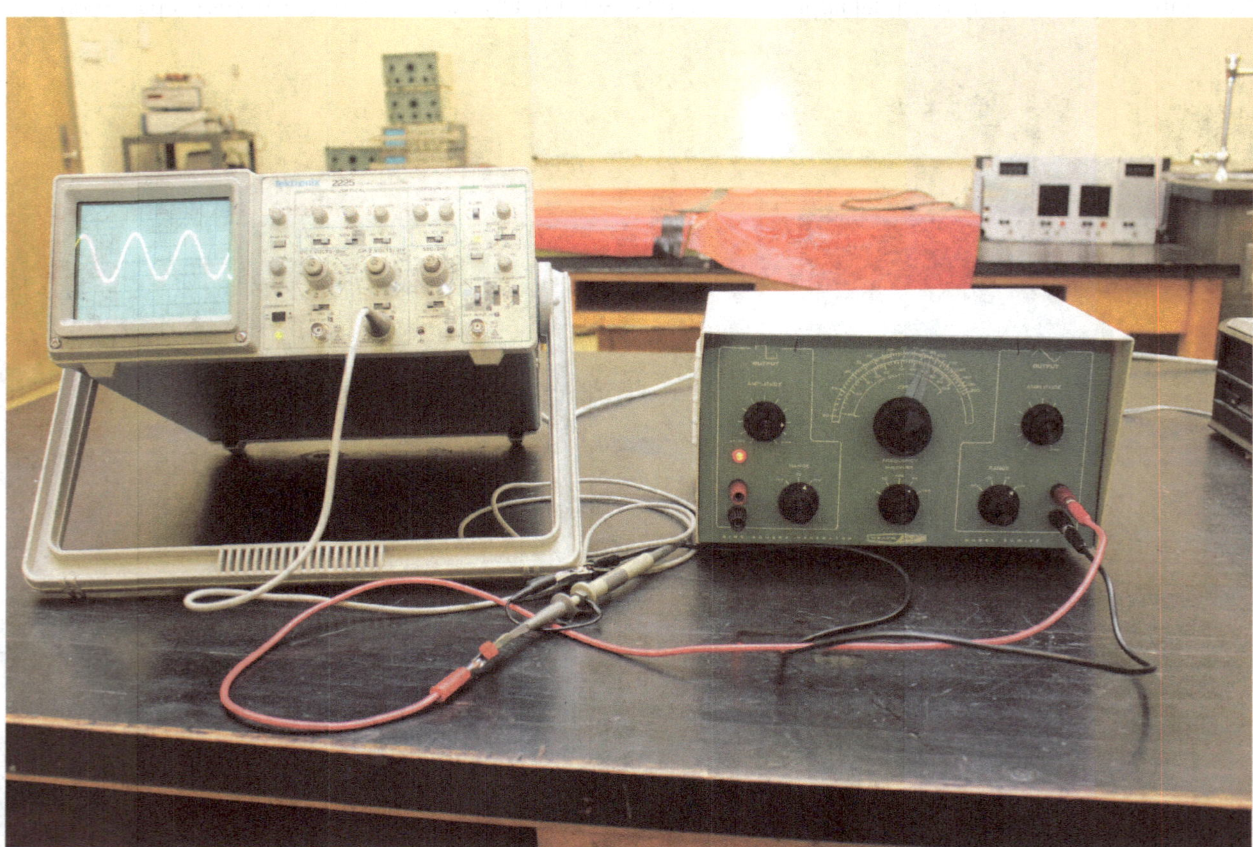

Figure 15.2   An oscilloscope displays the signal that it receives for a sine wave output terminal of an audio oscillator.

7. Adjust the horizontal position. What happened?
8. Adjust the amplitude of the audio oscillator. What happened?
9. Adjust the frequency of the audio oscillator. What happened?
10. Switch the output leads of the audio oscillator to the square wave.
11. Repeat Steps 4 through 9.
12. Disconnect the probe from the audio oscillator. Allow the negative lead to hang freely while touching the positive lead.
13. Adjust the oscilloscope trace until you see a signal. What do you see and why?

Data Record and Analysis

This laboratory exercise is observational and qualitative in nature. Record your responses to each question in the procedure in your laboratory notebook. In your laboratory report, you will repeat each question and record your response to it.

Accuracy of an Audio Oscillator

Purpose

To study the quantitative features of an oscilloscope and to determine the accuracy of an audio oscillator for several frequencies and voltage levels.

Theory

An audio oscillator produces a voltage changing at a frequency which would be audible if it were sent to a speaker. Usually it can be adjusted throughout the range 20 hertz (Hz) to 20,000 Hz or cycles per second, or more. You can also adjust the amplitude and choose either sine or square waves. All three of these adjustments have some error which you will check in this experiment.

Figure 16.1   A typical audio oscillator generates electrical signals whose frequencies are continuously variable between 20 Hz and 20,000 Hz. The audio oscillator produces both sine wave signals and square wave signals having continuously variable amplitude of four decades of potential difference.

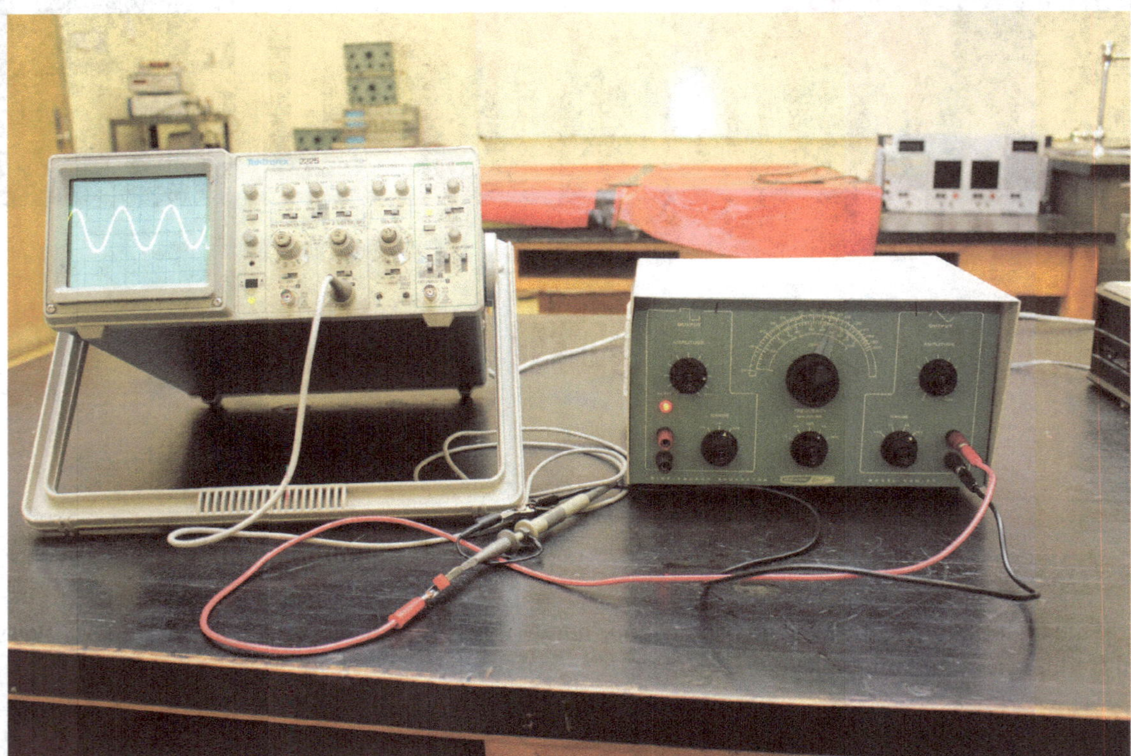

Figure 16.2   The electrical signal from the output terminal of the audio oscillator is displayed on the oscilloscope.

   Figure 16.1 shows a typical audio oscillator. An audio oscillator usually has a red and black output. The black is ground or zero volts. It should be connected to any ground connection on the oscilloscope. The red is the wave which you will display on the vertical input of the scope.

Equipment

Tektronix oscilloscope, audio oscillator, coaxial probe, and wires.

Procedure

1. Set your signal amplitude to 2 V.
2. Set your frequency to 20 Hz.
3. Adjust the VOLTS/DIV setting and SEC/DIV setting to get a readable trace. Record these values in the appropriate places in your data table.
4. Record the number of vertical divisions trough-to-crest and the number of horizontal divisions crest-to-crest.
5. Repeat Steps 2 through Step 4 for 50 Hz, 200 Hz, 500 Hz, 2000 Hz, 5000 Hz, and 20,000 Hz.
6. Repeat Steps 1 through Step 5 using signal amplitude setting of 5 V.
7. Switch to the square wave signal. Set the output to 5000 Hz and 5 V
8. Sketch the waveform.

Data Record and Analysis

Duplicate Table 16.1 in your laboratory notebook and laboratory report:

Table 16.1 Oscilloscope Data

f (Hz)	Amplitude (V) (Audio Oscillator)	VOLTS/DIV	DIVs (Vertical)	Amplitude (V) ½ VOLTS/DIV × DIVs	% error	SEC/DIV	DIVs (Horizontal)	T (sec) SEC/DIV × DIVs	$f = \dfrac{1}{T}$	% error
20	2									
50	2									
200	2									
500	2									
2000	2									
5000	2									
20000	2									
20	5									
50	5									
200	5									
500	5									
2000	5									
5000	5									
20000	5									

Reflection and Refraction

Purpose

To study the Law of Reflection. To study the Law of Refraction. To measure the index of refraction of lucite.

Theory

Law of Reflection. When a beam of light is incident on a surface as shown in Figure 17.1, its *angle of incidence* is measured from the normal as θ_i. The beam reflects from the surface at angle of reflection θ_r, which is also measured from the normal. The Law of Reflection states that the angle of reflection equals the angle of incidence, which may also be written in equation form:

$$\theta_r = \theta_i$$

Law of Refraction. When a beam of light is incident on a surface of a medium that is different than the incident medium, its path direction changes as shown in Figure 17.2. The direction change is called refraction. The change in direction is explained by the Law of Refraction which is also known as Snell's Law. Snell's Law in equation form is written as:

$$n_r \sin\theta_r = n_i \sin\theta_i$$

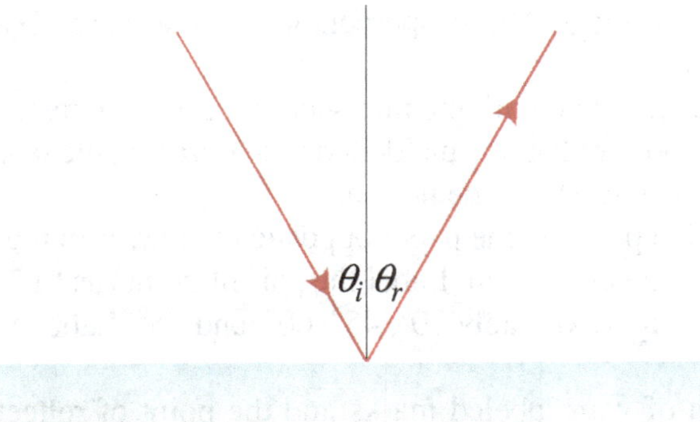

Figure 17.1 Reflection of a light ray from a smooth surface.

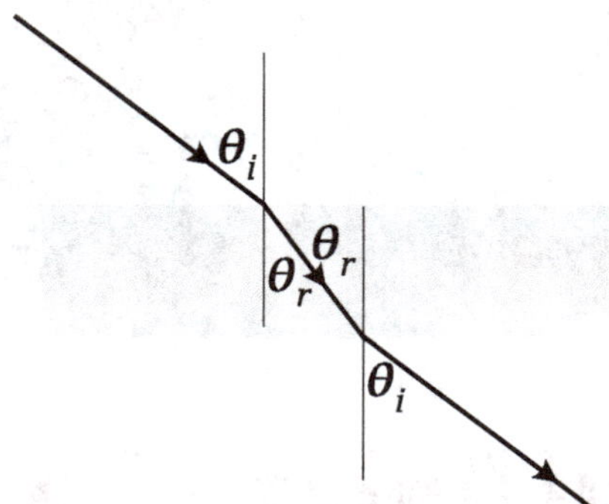

Figure 17.2   Refraction of a light ray at the interface of two media.

where n_r and n_i are the indices of refraction of the refracting and incident media, respectively, and θ_r and θ_i are the angles of refraction and incidence, respectively. The index of refraction of any medium is defined by the ratio of the speed of light in a vacuum to that in the medium or

$$n = \frac{speed\ of\ light\ in\ a\ vacuum}{speed\ of\ light\ in\ a\ medium} = \frac{c}{v}$$

The speed of light in a medium is always less than that its speed in a vacuum. Therefore, the index of refraction of every non-vacuum medium is greater than 1.0. As a practical matter, the speed of light in air is very close to the speed in a vacuum. Therefore, the index of refraction of air is assumed to have the same index as a vacuum, 1.0. For this experiment, the index of refraction of lucite can be determined using the following expression:

$$n = \frac{\sin\theta_i}{\sin\theta_r}$$

Equipment

Light box, mirror, lucite block, paper, protractor, and straight edge.

Procedure

1. Tape a sheet of paper to your lab tabletop. Use a straightedge to draw a line down the length of the paper lengthwise and near one edge. Draw a perpendicular line (normal) across the center of the page which intersects the first line.
2. Place your mirror along the first line with the intersection near its center.
3. Align the beam of the light box so that it is incident on the mirror at the intersection of the two lines and makes an angle of approximately 15° to the normal.
4. Use a pencil or pen to mark a point on the paper opposite the mirror in the center of the incident beam. Mark a similar point in the reflected beam. Label the pair of points as "1."
5. Repeat Step 3 and Step 4 at approximately 30°, 45°, 60°, and 75°. Label each pair as "2," "3," "4," and "5," respectively.
6. Draw a line between each of your labeled marks and the point of reflection. Use your protractor to measure each incident angle with respect to the normal. Record these values as θ_i in the appropriate location in your data table.
7. For each beam path: Measure the reflected angle with respect to the normal. Record its value as θ_r in the appropriate location in your data table.

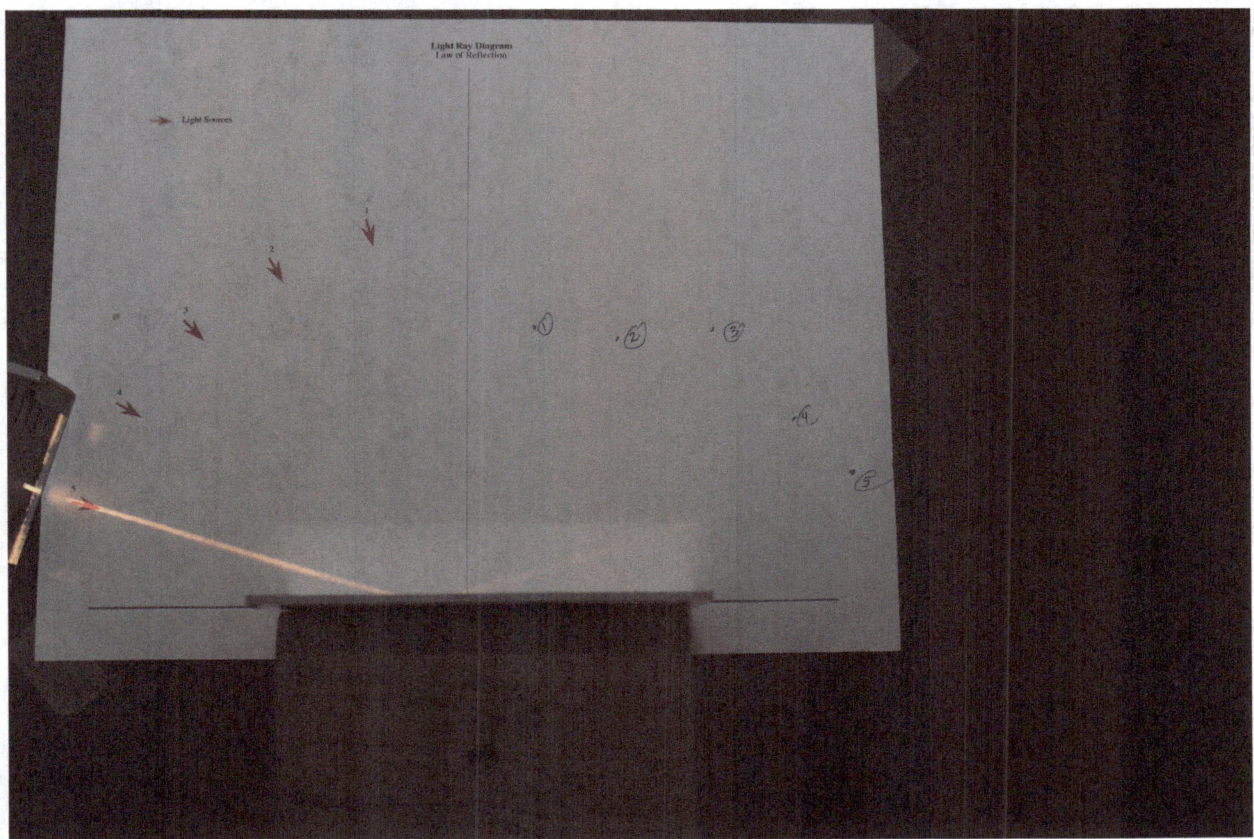

Figure 17.3   A beam of light reflects from the surface of a smooth plane mirror. The path of each reflected beam is marked so that they can be traced later.

8. Tape a new sheet of paper to your lab tabletop. Place the lucite block at the center of the paper in the same orientation as the paper. Trace the perimeter of the block.

9. Remove the block. Draw a normal line to the long side of the block which will be used as the incident side.

10. Carefully return the block to its position. Align the light beam so that it is incident on the normal intercept point at 15° to the normal. Use a pencil or pen to mark the center of the incident beam near the edge of the paper, the point where the beam emerges from the block, and a point in the center of the emerged beam near edge of the paper. Label the three marks as "1."

11. Repeat Step 11 for angles of 30°, 45°, 60°, and 75°. Label each set of marks as "2," "3," "4," and "5," respectively as shown in Figure 17.4.

12. Use your straightedge to connect the points of each set of marks. Each set includes the intersection of the normal and the side of the block.

13. *For each beam path*: Measure the angle between the incident beam and the normal. Record this value as $\theta_{air,1}$. Measure the angle between the emergent beam and the normal. Record this value as $\theta_{air,2}$. Measure the angle between the refracted beam and the normal. Record this value as $\theta_{lucite,1}$. Measure the angle between the normal and the refracted beam just before it emerges from the block. Record this value as $\theta_{lucite,2}$ as shown in Figure 17.5.

14. On Graph 1, plot each pair of angular data points with $\theta_{air,ave}$. on the vertical axis and $\theta_{lucite,ave}$. on the horizontal axis. Draw a smooth curve through the points.

15. Calculate the sines of $\theta_{air,ave}$ and $\theta_{lucite,ave}$ for each beam. On Graph 2, plot $\sin\theta_{air,ave}$ versus $\sin\theta_{lucite,ave}$ for each line. Draw the best fit straight line through the points.

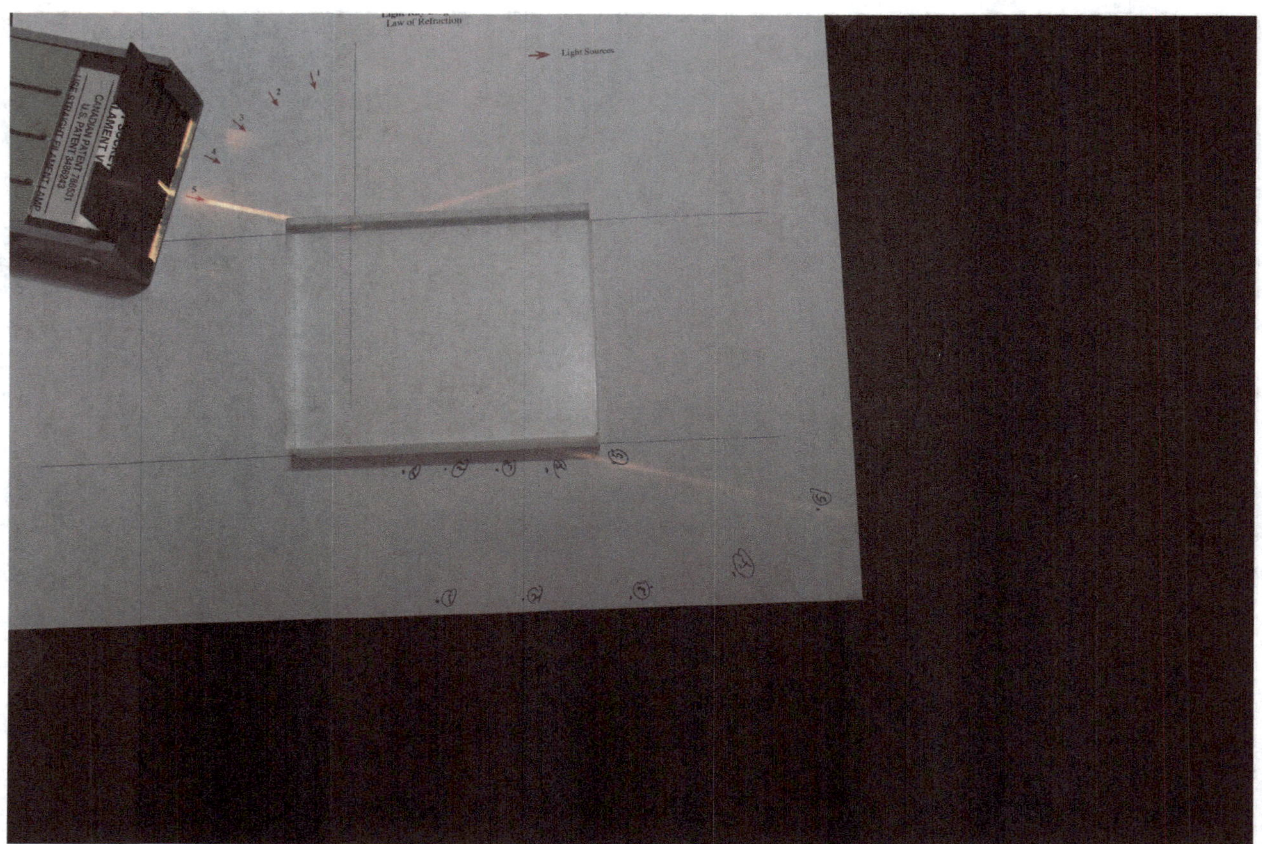

Figure 17.4   A beam of light refracts through a transparent block.

Data Record and Analysis

Duplicate Tables 17.1 and 17.2 in your laboratory notebook and laboratory report:

Table 17.1   Reflection Data

Beam	θ_i	θ_r	$\%error = \dfrac{\theta_r - \theta_i}{\theta_i} \times 100\%$
1			
2			
3			
4			
5			

$n_{\text{std}} = 1.49$

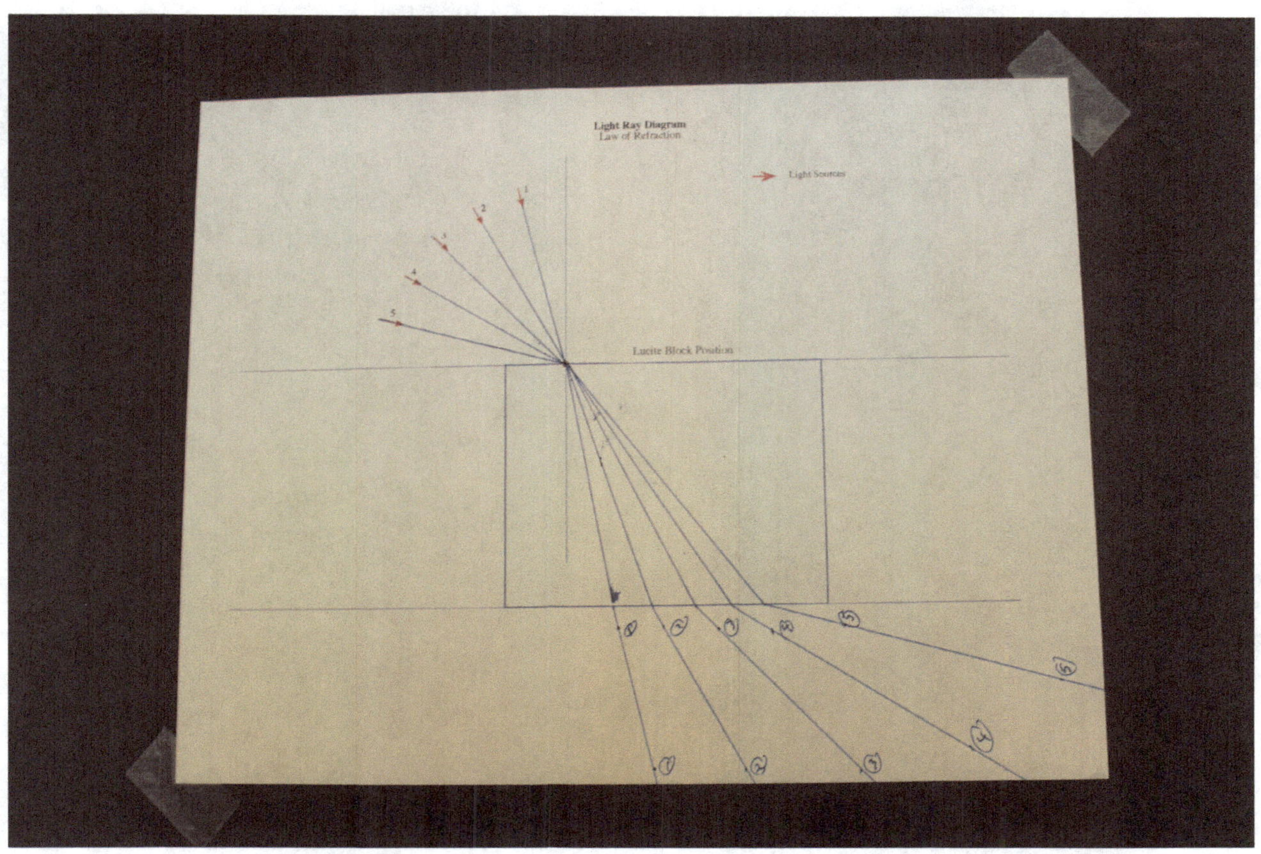

Figure 17.5 The paths of five refracted light beams are traced on white paper.

Table 17.2 Refraction Data

Beam	θ_{air} 1	2	ave.	θ_{lucite} 1	2	ave.	n
1							
2							
3							
4							
5							
ave.							

$$\%\,error = \frac{n_{ave} - n_{std}}{n_{std}} \times 100\% = \underline{\hspace{2cm}}$$

$$n_{graph} = \text{Graph 2 slope} = \underline{\hspace{3cm}}$$

$$\%\,error = \frac{n_{graph} - n_{std}}{n_{std}} \times 100\% = \underline{\hspace{2cm}}$$

Mirrors and Lenses

Purpose

To study the optical properties of convex lenses and concave mirrors.

Theory

Concave Mirrors. When parallel light rays reflect from the surface of a concave mirror, the rays converge to a point called the focus, F, as shown in Figure 18.1.

Light from a real object may reflect from the surface of a concave mirror to form an image on a screen. This case is illustrated by the ray diagram in Figure 18.2.

The location of the image may be calculated using the mirror equation:

$$\frac{1}{f} = \frac{1}{s} = \frac{1}{s'}$$

where f is the focal length of the mirror, s is the object distance, and s' is the image distance. The magnification of the image is defined as the ratio of the image height to the object height:

$$M = \frac{h'}{h} = -\frac{s'}{s}$$

where M is the magnification, h' is the image height, and h is the object height.

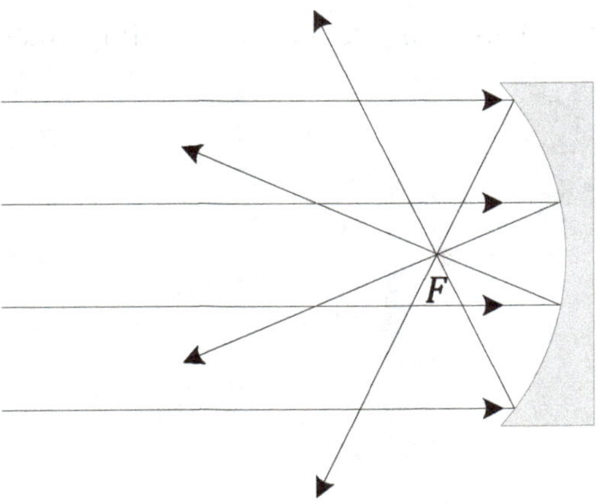

Figure 18.1 Parallel rays converge at focal point F after reflection on the surface of a concave mirror.

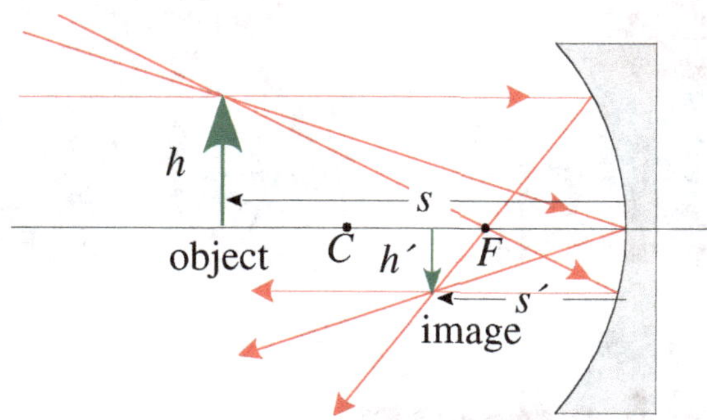

Figure 18.2 Light rays from an object refract through a double-convex lens and converge to form a real image.

Convex Lenses. When parallel light rays refract through a convex lens, the refracted rays converge to a point called the focus, F, as shown in Figure 18.3.

Light from a real object may refract through a convex lens to form a real image on the screen. This case is illustrated by the ray diagram in Figure 18.4.

The location of the image may be calculated using the lens equation:

$$\frac{1}{f} = \frac{1}{s} + \frac{1}{s'}$$

where f is the focal length of the lens, s is the object distance, and s' is the image distance. The magnification of the image is defined as the ratio of the image height to the object height:

$$M = \frac{h'}{h} = -\frac{s'}{s}$$

as in the case of mirrors.

In this laboratory session, the images will be focused on a screen. Images of this type are known as *real images*. Both object distance s and image distance s' are positive. Real images produced in this way are inverted or upside down. They have negative magnifications.

There are also images which can be viewed only by sight. They cannot be focused on a screen. Images of this type are known as *virtual images*. In the case of virtual images, s' is negative. Virtual images produced in this way are erect or right side up. They have positive magnifications.

Equipment

Optical bench, light source, convex lens, concave mirror, mounting brackets, lens holder, screen holder, ruler, and incandescent light.

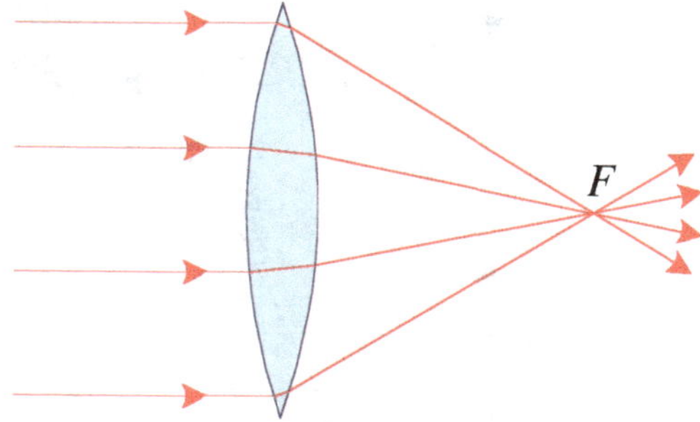

Figure 18.3 Parallel rays refract through a double-convex lens and converge at focal point F.

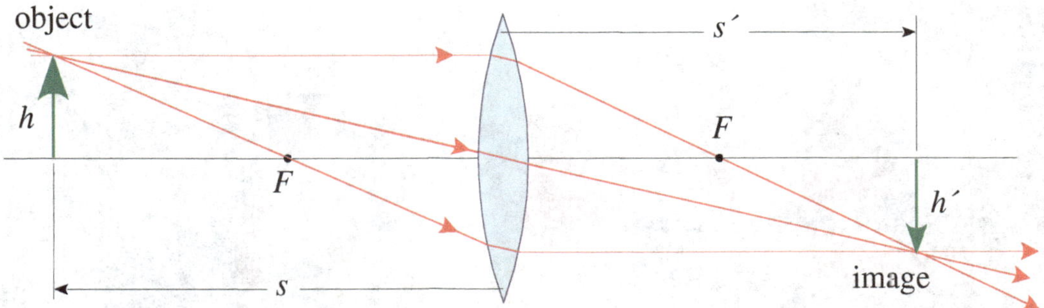

Figure 18.4   Light rays from an object refract through a double-convex lens and converge to form a real image.

Procedure

1. Set up the optical bench with the light source near one end, the screen on the other, and the lens in the middle as shown in Figure 18.5.
2. Plug in the light source. Adjust the lens and screen position until the image is focused on the screen.
3. Record the position of the screen (x_{screen}), the position of the lens (x_{lens}), and the position of the source (x_{source}).
4. Use the ruler to measure the length of the arrow on the source. Record its length as h. Measure the length of the image of the arrow on the screen. Record its length as h'. Because the image is inverted, h' should be recorded as a negative value.

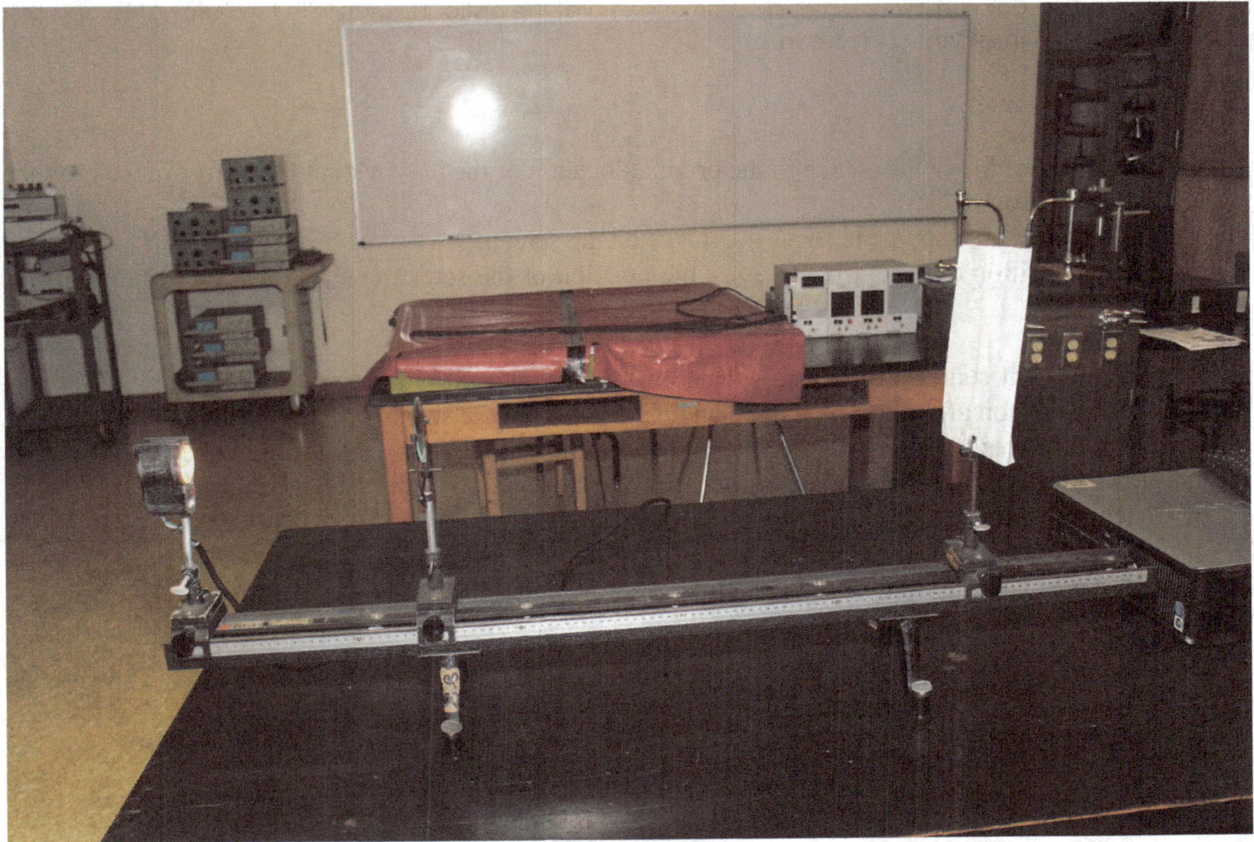

Figure 18.5   The lens is mounted on the optical between between the illuminated object and the screen.

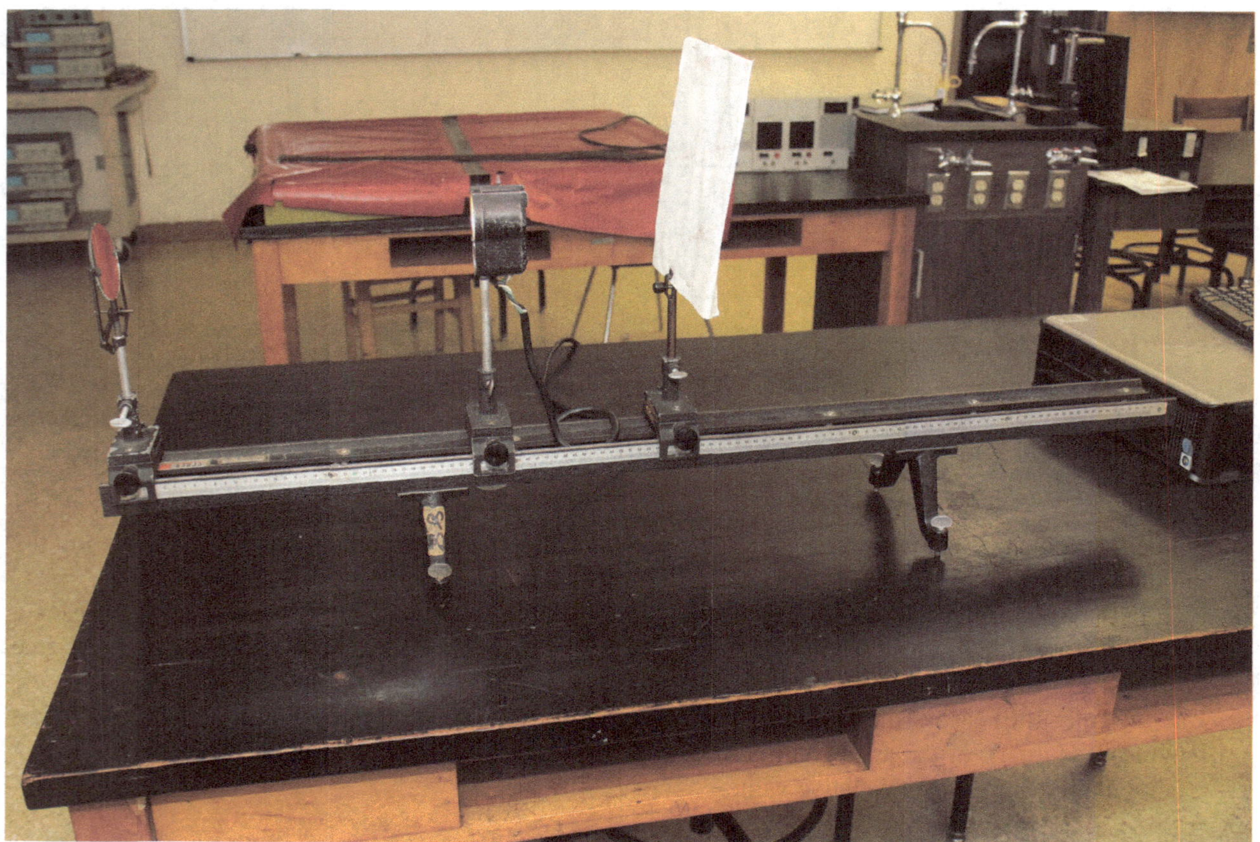

Figure 18.6   The concave mirror is mounted on one end of the optical bench with the illuminated object mounted between the mirror and screen.

5. Remove the lens. Move the source to the original location of the lens. Mount the mirror in the original location of the source.
6. Adjust the screen and source to focus the image on the screen.
7. Record the position of the mirror (x_{mirror}), the position of the screen (x_{screen}), and the position of the source (x_{source}).
8. Repeat Step 4.
9. Remove the light source. Adjust the screen until a distant light is focused to a point on the screen. Record the position of the mirror (x_{mirror}) and the position of the screen (x_{screen}).
10. Exchange the screen and lens holder. Replace the mirror with the lens. Adjust the screen until the distant light is focused to a point on the screen. Record the position of the screen (x_{screen}) and the position of the lens (x_{lens}).

Data Record and Analysis

Duplicate Table 18.1 in your laboratory notebook and laboratory report:

Table 18.1   Lens and Mirror Data

Lens		Mirror	
x_{screen}		x_{screen}	
x_{lens}		x_{mirror}	
x_{source}		x_{source}	
h		h	
h'	$(-)$	h'	$(-)$
$s = x_{lens} - x_{source}$		$s = x_{source} - x_{mirror}$	
$s' = x_{screen} - x_{lens}$		$s' = x_{screen} - x_{mirror}$	
$f_{exp} = \dfrac{ss'}{s + s'}$		$f_{exp} = \dfrac{ss'}{s + s'}$	
$M_{exp} = -\dfrac{s'}{s}$		$M_{exp} = -\dfrac{s'}{s}$	
$M_{std} = -\dfrac{h'}{h}$		$M_{std} = -\dfrac{h'}{h}$	
$\%error\ (f)$		$\%error\ (f)$	
$\%error\ (M)$		$\%error\ (M)$	
Distant Object			
x_{lens}		x_{mirror}	
x_{screen}		x_{screen}	
$f_{std} = x_{lens} - x_{screen}$		$f_{std} = x_{screen} - x_{mirror}$	

Microscope

Purpose

To build a model microscope and to study its features.

Theory

A simple microscope is composed of two convex lenses, a very short focal length objective and a longer focal length eyepiece. The object is placed near the focal point of the objective lens. A real image is formed between the objective and eyepiece. The eyepiece forms a virtual image 25 cm away. The microscope is diagramed in Figure 19.1:

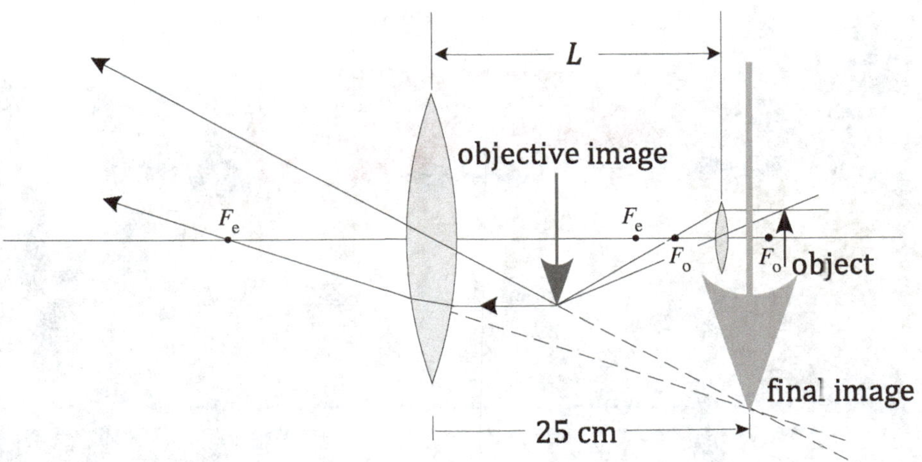

Figure 19.1 Diagram of a model microscope.

The object distance of the objective (s_o) is approximately equal to its focal length (f_o). Its image distance (s_o') is approximately equal to the tube length of the microscope (L). Therefore, the magnification of the microscope objective is given by:

$$M_o = -\frac{s_o'}{s_o} \cong -\frac{L}{f_o}$$

The object distance of the eyepiece (s_e) is approximately equal to its focal length (f_e). The image distance of the eyepiece (s_e') is adjusted to the near point for normal vision (–25 cm). The image distance is negative because the eyepiece image is virtual. The magnification of the microscope eyepiece is given by:

$$M_e = -\frac{s_e'}{s_e} \cong \frac{25 \text{ cm}}{f_e}$$

The total magnification of the microscope is the product of the magnifications of the objective and eyepiece:

$$M = M_o M_e = -\frac{L}{f_o}\frac{25 \text{ cm}}{f_e}$$

Equipment

Optical bench, light source, convex lenses, mounting brackets, lens holders, screen holder, ruler, and magnification paper strip.

Procedure

1. Set of the light source at one end of the optical bench ($x_{source} = 0.0$ cm). Set the objective lens approximately 6 cm to 7 cm from the light source as shown in Figure 19.2.
2. Move the screen until a magnified image is clearly focused on the screen. Record the positions of the objective lens ($x_{objective}$) and the screen (x_{screen}).

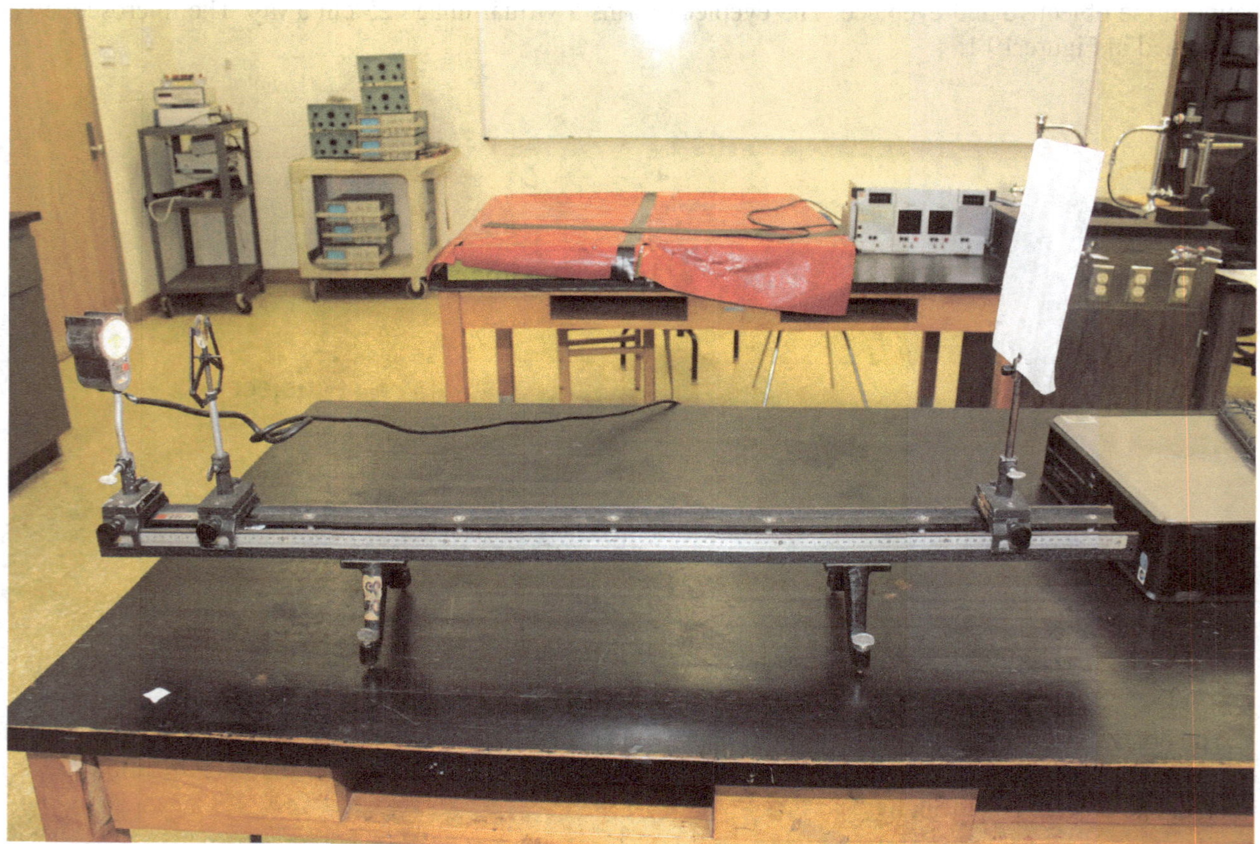

Figure 19.2   The illuminated object on one end of the optical bench is mounted near the focal point of the objective lens. A real image is formed on a screen mounted near the other end of the optical bench.

Figure 19.3 The model microscope is constructed using a small focal length objective lens and a long focal length eyepiece lens. The model microscope is used to view the illuminated object.

3. Replace the screen with the eyepiece lens as shown in Figure 19.3.
4. Move the eyepiece lens until a virtual image is in clear focus when you look through the eyepiece lens. Record the position of the eyepiece lens ($x_{eyepiece}$).
5. Tape the calibration strip to the source. Use the ruler to measure the distance between adjacent lines in the image on the eyepiece. Record this value as h'. The distance between lines on the strip is $h = 1.0$ mm.

Data Record and Analysis

Duplicate the following data and calculations in your laboratory notebook and laboratory report:

$x_{source} = $ _______________________________

$x_{objective} = $ _______________________________

$x_{screen} = $ _______________________________

$x_{eyepiece} = $ _______________________________

$h = 1.0$ mm _______________________________

$h' = $ _______________________________

$M_{estimate} = \dfrac{h'}{h} = $ _______________________________

$$L = x_{\text{eyepiece}} - x_{\text{objective}} = \underline{\hspace{4cm}}$$

$$s_o = x_{\text{objective}} - x_{\text{source}} = \underline{\hspace{4cm}}$$

$$s_o' = x_{\text{screen}} - x_{\text{objective}} = \underline{\hspace{4cm}}$$

$$s_e = x_{\text{eyepiece}} - x_{\text{screen}} = \underline{\hspace{4cm}}$$

$$s_e' = -25 \text{ cm}$$

$$f_o = \frac{s_o s_o'}{s_o + s_o'} = \underline{\hspace{4cm}}$$

$$f_e = \frac{s_e s_e'}{s_e + s_e'} = \underline{\hspace{4cm}}$$

$$M_o = -\frac{s_o'}{s_o} = \underline{\hspace{4cm}}$$

$$M_e = -\frac{s_e'}{s_e} = \underline{\hspace{4cm}}$$

$$M_{\text{exp}} = M_o M_e = \underline{\hspace{4cm}}$$

$$M_{\text{std}} = -\frac{L(25 \text{ cm})}{f_o f_e} = \underline{\hspace{4cm}}$$

$$\%error_{\text{exp}} = \frac{M_{\text{exp}} - M_{\text{std}}}{M_{\text{std}}} = \underline{\hspace{4cm}}$$

$$\%error_{\text{est}} = \frac{M_{\text{est}} - M_{\text{std}}}{M_{\text{std}}} = \underline{\hspace{4cm}}$$

Interference

Purpose

To study the interference pattern of monochromatic light through a double-slit.

Theory

When light waves (and waves in general) occupy the same space, they interfere with each other. Their interference may be constructive if the waves enhance each other, or destructive if the waves diminish each other. In the case of coherent waves, such as laser light, the resulting interference forms a pattern of bright spots as shown in Figure 20.1 where the light passing through the slits constructively interfere. This pair of slits is known properly as Young's double-slit.

The interference due to the double-slit is diagrammed in Figure 20.2.

Light from the two slits arrive at the screen with a phase difference due to their difference in path. The path difference is given by:

$$\delta = d \sin\theta$$

where d is the slit separation and is the angular displacement between the incident laser beam and the location on the screen. If the path difference is equal to an integral or whole number of wavelengths, then the two beams are in phase. Constructive interference yields bright spots if path difference $\delta = n\lambda$ for wavelength λ and $n = 1, 2, 3, \ldots$. The locations of the bright spots are given by:

$$\delta = n\lambda = d \sin\theta$$

In the case of the double-slit in this experiment, $\sin\theta = x/D$ where x is the distance from the central bright spot and D is the distance from the slits to the screen. Using the more easily measured values, the locations of bright spots may also be determined by the equation:

$$n\lambda = \frac{dx}{D}$$

The bright spots are evenly spaced. The distance Δx between any two spots is determined by Δn, their difference in n:

$$\Delta n\lambda = \frac{d\Delta x}{D}$$

Figure 20.1   An interference pattern is formed on a screen by a laser beam that has passed through a pair of slits.

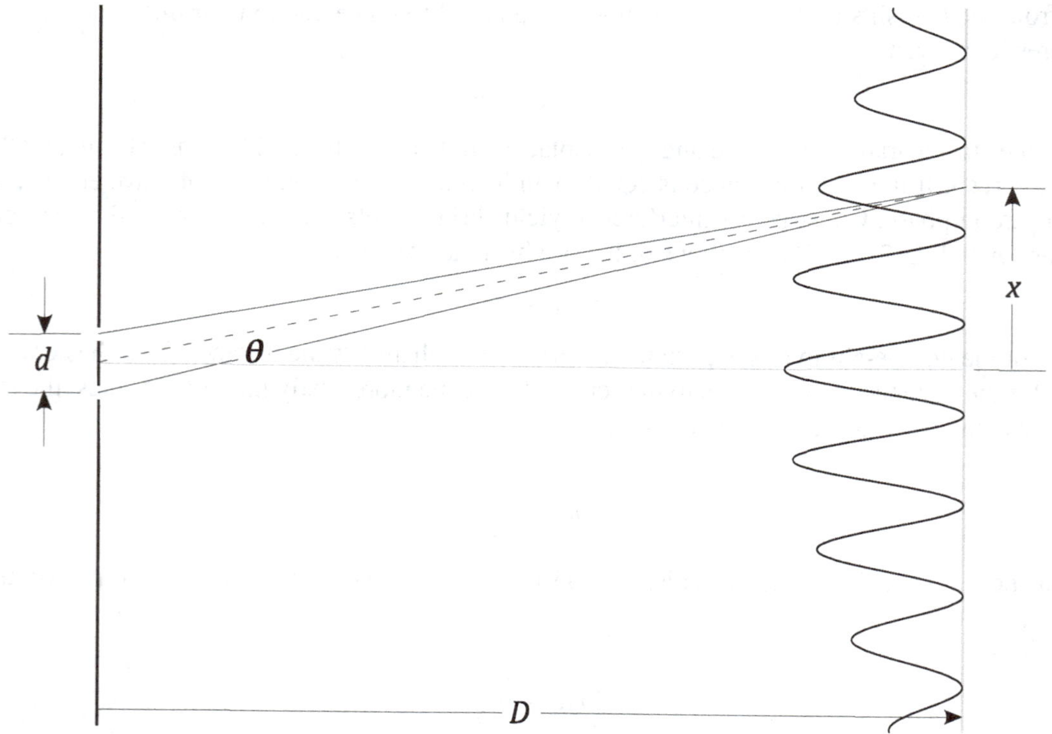

Figure 20.2   Illustration of the intensity of laser beam after it passes through Young's double-slit.

In the case of adjacent bright spots, $\Delta n = 1$. The wavelength of the laser may be determined from the previous expression:

$$\lambda = \frac{d\Delta x}{D}$$

Equipment

Helium-neon laser, slits, screen mount, tape measure, ruler, wooden blocks, and adhesive tape.

Procedure

1. Setup the apparatus so that a pattern of spots is visible on the screen while the laser is in operation.
2. Tape a sheet of white paper on the screen mount so that the spot pattern appears on the paper.
3. Select five (5) consecutive bright spots. Using a ruler, measure the distance between the centers of the spots on either end of your selection. Record this value as x.
4. Measure the distance between the slits and the screen. Record this value as D.

Data Record and Analysis

Duplicate the following data and calculations in your laboratory notebook and laboratory report:

$d = 2.0 \times 10^{-4}\,\text{m}$

$x = \underline{\hspace{10cm}}$

$D = \underline{\hspace{10cm}}$

$\Delta x = \dfrac{x}{4} = \underline{\hspace{8cm}}$

$\lambda_{std} = 632.8\,\text{nm}$

$\lambda_{exp} = \dfrac{d\Delta x}{D} = \underline{\hspace{8cm}}$

$@error = \dfrac{\lambda_{exp} - \lambda_{std}}{\lambda_{std}} \times 100\% = \underline{\hspace{4cm}}$